PMP®EXAM
CHALLENGE!

Sixth Edition

PMP® EXAM CHALLENGE!

Sixth Edition

J. LeRoy Ward, PMP, PgMP
Ginger Levin, PMP, PgMP

CRC Press
Taylor & Francis Group
Boca Raton   London   New York

CRC Press is an imprint of the
Taylor & Francis Group, an **informa** business
AN AUERBACH BOOK

CRC Press
Taylor & Francis Group
6000 Broken Sound Parkway NW, Suite 300
Boca Raton, FL 33487-2742

First issued in hardback 2017

© 2014 by Taylor & Francis Group, LLC
CRC Press is an imprint of Taylor & Francis Group, an Informa business

No claim to original U.S. Government works

ISBN 13: 978-1-138-44031-9 (hbk)
ISBN 13: 978-1-4665-9982-6 (pbk)

Library of Congress Cataloging-in-Publication Data

Ward, J. LeRoy.
    [PMP challenge]
    PMP exam challenge! / Ginger Levin, J. LeRoy Ward. -- Sixth edition.
        pages cm. -- (ESI International project management series)
    Revised edition of: PMP exam challenge! / J. LeRoy Ward, Ginger Levin. 5th ed.
    Includes bibliographical references and index.
    ISBN 978-1-4665-9982-6 (pbk. : alk. paper) 1. Project management--Examinations, questions, etc. I. Levin, Ginger. II. Title.

HD69.P75W37 2013
658.4'04--dc23
                                                            2013013393

Visit the Taylor & Francis Web site at
http://www.taylorandfrancis.com

and the CRC Press Web site at
http://www.crcpress.com

Contents

Preface

About the Authors

Acknowledgments

Introduction

PREFACE

ESI has been teaching people how to prepare for the PMP® since January 1991, using our popular and proven *PMP® Exam Preparation* course to thousands of people who tell us it works Using information from this course and *A Guide to the Project Management Body of Knowledge (PMBOK® Guide)*—Fifth Edition, 2013, we developed the ***PMP Challenge!*** to assist you as you prepare for this exam. We developed 60 questions for each of the ten knowledge areas, or 600 questions, which will challenge you to examine the *PMBOK® Guide* at the level of detail required to dramatically increase your probability of passing the exam. We also included references to the five project management process groups—the focus of the PMP® exam. Study hard—it is worth it. Good luck in pursuing your PMP®.

We also believe that this book is a valuable reference for those of you who are pursuing the CAPM®. Study hard, and we wish you success in attaining the CAPM®.

Ginger Levin, DPA, PMP, PgMP
Lighthouse Point, Florida

J. LeRoy Ward, PMP, PgMP
New York, New York

About the Authors

Dr. Ginger Levin, PMP, PgMP, and a certified *OPM3* Professional is a consultant and educator in portfolio, program, and project management and has conducted numerous maturity assessments over the past 20 years. She has 47 years of experience working in the private sector, the U.S. Federal Government, and in consulting and training. She is an Adjunct Professor at the master's degree level in project management for the University of Wisconsin-Platteville and at the doctoral level for SKEMA University in Lille, France and RMIT in Melbourne, Australia. She holds a doctorate in public administration from The George Washington University, where she also received the outstanding dissertation award. Dr. Levin is the editor, author, or co-author of 14 books and is an active member of the Project Management Institute.

J. LeRoy Ward, PMP, PgMP, and CSM is the Executive Vice President of ESI International, where he is the principal executive responsible for R&D, Product Strategy, Consulting, and Corporate Marketing. ESI International is the world's largest project-focused training company with curriculums addressing project and program management, business analysis, contract management, sourcing management, and business skills. A PMP® since 1990, Mr. Ward developed ESI's popular *PMP Exam Preparation* course and taught it and

ABOUT THE AUTHORS CONTINUED

other courses to people from more than 50 countries. He has almost 40 years of experience in the public and private sectors and is a popular and dynamic speaker in the field. He holds B.S. and M.S. degrees from Sothern Connecticut University and an M.S.T.M. degree with distinction from The American University.

ACKNOWLEDGMENTS

The production of a specialty publication requires the contributions of many people, and we would like to thank each of those who helped prepare the fifth edition of *PMP® Exam Challenge!*

Our special thanks to Sara Halgas of ESI International and the following people at CRC:

John Wyzalek, Randy Burling, James Yanchak, Theresa Gutierrez and her team, York Lambrecht and his team, and Christopher Manion.

INTRODUCTION

Preparing for and passing the PMP® exam is no small feat. Even though the number of certified PMP®s continues to grow at a phenomenal rate, the failure rate also remains high. **PMP® Exam Challenge!** has been designed to help you pass the exam in an easy-to-use, highly portable publication, containing the key relevant topics you may encounter on the exam. And, in this Sixth Edition, we have updated the content of the ten knowledge areas to reflect the *PMBOK® Guide,* Fifth Edition.

Featuring an entirely new section on Program Stakeholder Management, **PMP® Exam Challenge!** uses a flash-card format to help you drill through the essential points of the *PMBOK® Guide.* The pages are tabbed so you can directly go to the knowledge area you need to study. Each page has plenty of white space that you may use to write notes.

Carry, and, of course, use **PMP® Exam Challenge!** everywhere you go during the days leading up to the exam. Start with any section. Proceed to a new section only after you have mastered the one you are currently studying. Use this book again and again. Work with your spouse, a friend, or colleague. Prior to the exam, do not allow a day pass without studying the questions and answers.

INTRODUCTION CONTINUED

Note that answers list the five project management process groups— Initiating, Planning, Executing, Monitoring and Controlling, and Closing. These references help you to answer questions that test your knowledge of when certain events or actions take place during the project management life cycle.

We hope this book makes drilling questions and answers fun. It is hard work, but earning the PMP® is a prestigous achievement— one that contributes to your professional growth in many ways. Whether you have been in project management for 30 years or 30 days, you will learn a lot by studying **PMP® Exam Challenge!**

PROJECT **Integration** MANAGEMENT

1

QUESTION

Define Project Integration
Management.

2

QUESTION

What six major processes are
included in Project Integration
Management?

ANSWER

The processes required to ensure that the various project elements are identified, combined, unified, and coordinated effectively.

[Initiating, Planning, Executing, Monitoring and Controlling, and Closing]

PROJECT **Integration** MANAGEMENT

ANSWER

Develop project charter, develop project management plan, direct and manage project work, monitor and control project work; perform integrated change control; and close project or phase.

[Initiating, Planning, Executing, Monitoring and Controlling, and Closing]

PROJECT **Integration** MANAGEMENT

3 QUESTION

List the one initiating process
in Project Integration
Management.

4 QUESTION

What is the purpose of the
direct and manage project
work process?

ANSWER

Develop project charter.

[Initiating]

ANSWER

The process of leading and performing the work in the project management pan and implementing approved changes to achieve project objectives.

[Executing]

5

What is meant by integrated
change control?

6

What are two tools and
techniques used to develop
the project management plan?

ANSWER

A formal process to review, approve, and manage changes to project deliverables, documents, organizational process assets, and the project management plan throughout the project and communicating their disposition.

[Monitoring and Controlling]

PROJECT **Integration** MANAGEMENT

ANSWER

- Expert judgment.
- Facilitation techniques.

[Planning]

PROJECT **Integration** MANAGEMENT

7
QUESTION

What is a major factor in
making and prioritizing
project selection decisions?

8
QUESTION

What are the four inputs to the
develop project management
plan process?

ANSWER

The performing organization's strategic plan.

[Initiating]

PROJECT **Integration** MANAGEMENT

ANSWER

- Project charter
- Outputs from planning processes
- Enterprise environmental factors
- Organizational process assets

[Planning]

PROJECT **Integration** MANAGEMENT

9

What are the three categories of organizational process assets used in developing the project charter?

10

Provide five examples of the project management information system (PMIS) that can be used in direct and manage project work.

ANSWER

- Organizational standards, processes, policies, and process definitions.

- Templates.

- Historical information and lessons learned knowledge base.

[Initiating]

ANSWER

- Scheduling software

- Configuration management system

- Information collection and distribution system

- Web interface to other online automated systems

- Work authorization system

[Executing]

11

QUESTION

What is the purpose of the
project charter?

12

QUESTION

In the close project process,
what are the three updates to
organizational process assets?

ANSWER

To authorize a project or a phase. It also documents the business needs, current understanding of the customer's needs and high-level requirements, assumptions and constraints, and the requirements for the project's new product, service, or result. It provides the project manager with the authority to apply organizational resources to the project.

[Initiating]

ANSWER

- Project files
- Project or phase closure documents
- Historical information

[Closing]

13

When should the project manager be identified and assigned?

14

What are three tools and techniques used in close project or phase?

ANSWER

As early in the project as is feasible, but always prior to the start of project planning and preferably during project charter development.

[Initiating]

ANSWER

- Expert judgment
- Analytical techniques
- Meetings

15

QUESTION

What are the three items referenced in the project statement of work?

16

QUESTION

Who should issue the project charter?

Aɴꜱᴡᴇʀ

- Business need
- Product scope description
- Strategic plan

[Initiating]

Aɴꜱᴡᴇʀ

The sponsor, PMO staff person, or a portfolio governing body chairperson, or an authorized representative at a level appropriate to procure funding and commit resources to the project.

[Initiating]

17

QUESTION

What is a project management plan?

18

QUESTION

What does the project charter authorize the project manager to do?

ANSWER

A formal, approved document that defines how a project is executed, monitored and controlled, and closed.

[Planning]

ANSWER

Authorizes the project manager to apply resources to project activities.

[Initiating]

19 QUESTION

During which project integration management process is the majority of the project's budget spent?

20 QUESTION

What is meant by corrective actions?

ANSWER

Direct and manage project work

[Executing]

ANSWER

An intentional activity that realigns performance of the project work with the project management plan

[Executing and Monitoring and Controlling]

21

What is involved in the close project process?

22

What are the three tools and techniques for direct and manage project work?

ANSWER

Finalizing all activities across all Project Management Process Groups to formally complete the phase or project

[Closing]

PROJECT **Integration** MANAGEMENT

ANSWER

- Expert judgment
- Project management information system
- Meetings

[Executing]

PROJECT **Integration** MANAGEMENT

23

QUESTION

What is a deliverable?

24

QUESTION

What are the five outputs from the direct and manage project work process?

ANSWER

Any unique and verifiable product, result, or capability to perform a service that must be produced and provided to complete a process, phase, or the project itself.

[Executing]

PROJECT **Integration** MANAGEMENT

ANSWER

- Deliverables
- Work performance data
- Change requests
- Project management plan updates
- Project document updates

[Executing]

PROJECT **Integration** MANAGEMENT

25 QUESTION

Why is change control
necessary on projects?

26 QUESTION

What are three types of
meetings?

ANSWER

Because projects rarely run exactly according to the project management plan.

[Monitoring and Controlling]

ANSWER

- Information exchange

- Brainstorming, option evaluation, or design

- Decision making

[Executing and Monitoring and Controlling]

27

QUESTION

List three configuration management activities included in the integrated change control process.

28

QUESTION

What should be done if the project is terminated prior to completion?

ANSWER

- Configuration identification
- Configuration status accounting
- Configuration verification and audit

[Monitoring and Controlling]

ANSWER

Document the reasons for actions taken if the project is terminated before completion.

[Closing]

29

QUESTION

What is meant by a defect repair?

30

QUESTION

What is an element of the project management plan that affects the perform integrated change control process?

ANSWER

An intentional activity to modify a nonconforming product or product component

[Executing and Monitoring and Controlling]

PROJECT **Integration** MANAGEMENT

ANSWER

Change management plan as it provides direction for managing the change control process and documents the formal change control board

[Planning and Monitoring and Controlling]

PROJECT **Integration** MANAGEMENT

31

QUESTION

What are the seven inputs
to the Monitor and Control
Project Work process?

32

QUESTION

What is a change control
board (CCB)?

ANSWER

- Project management plan
- Schedule forecasts
- Cost forecasts
- Validated changes
- Work performance information
- Enterprise environmental factors
- Organizational process assets

[Monitoring and Controlling]

PROJECT **Integration** MANAGEMENT

ANSWER

A group of people responsible for reviewing, evaluating, approving, delaying, or rejecting changes to a project and recording and communicating the decisions

[Monitoring and Controlling]

PROJECT **Integration** MANAGEMENT

33

What three items are part of the project approval requirements in the project charter?

34

What are the four inputs to perform integrated change control?

Aₙₛwₑᵣ

- What constitutes project success

- Who decides if the project is a success

- Who signs off on the project

[Initiating]

Aₙₛwₑᵣ

- Approved change requests

- Change log

- Project management plan updates

- Project document updates

[Monitoring and Controlling]

35

QUESTION

List seven sources of expert judgment.

36

QUESTION

How is expert judgment used in perform integrated change control?

ANSWER

- Other units in the organization
- Consultants
- Stakeholders, including customers or sponsors
- Professional and technical associations
- Industry groups
- Subject matter experts
- Project management office

[Initiating]

ANSWER

As a tool and technique. Experts serve on the change control board to approve, review, evaluate, delay, or reject changes to the project baselines.

[Monitoring and Controlling]

37

List seven reasons for which a business case for a project may be created.

38

What is the primary objective of the business case?

ANSWER

- Market demand
- Organizational need
- Customer request
- Technological advance
- Legal requirement
- Ecological impact
- Social need

[Initiating]

ANSWER

To help the organization determine if the project is worth the required investment.

[Initiating]

39

What are three elements of the project management plan that may be used in integrated change control?

40

List three items that may be part of work performance information.

ANSWER

- Scope management plan
- Scope baseline
- Change management plan

[Planning and Monitoring and Controlling]

PROJECT **Integration** MANAGEMENT

ANSWER

- Status of deliverables
- Implementation status for change requests
- Forecasted estimate to complete

[Executing and Monitoring and Controlling]

PROJECT **Integration** MANAGEMENT

41

QUESTION

What is the difference between a standard and a regulation?

42

QUESTION

What is a change control system?

Answer

A standard provides for common and repeated use, rules, guidelines, or characteristics for activities or results. A regulation is a government-imposed requirement for which compliance is mandatory.

[Planning and Executing]

Answer

A set of procedures that describes how modifications to the project deliverables and documentation are managed and controlled

[Monitoring and Controlling]

43 QUESTION

Change requests are issued,
and approved changes
are implemented in which
process?

44 QUESTION

What are approved change
requests?

A NSWER

Direct and manage project work

[Executing]

A NSWER

A change request that has been processed through the integrated change control process and approved

[Executing and Monitoring and Controlling]

45

QUESTION

Why is a change log used?

46

QUESTION

What are three examples of personnel administration as an enterprise environmental factor in the direct and manage project work process?

ANSWER

To document changes that occur during a project especially in terms of time, cost, and risk; it also includes rejected change requests

[Monitoring and Controlling]

ANSWER

- Hiring and firing guidelines
- Employee performance reviews
- Training records

[Executing]

47 QUESTION

What is rolling wave planning?

48 QUESTION

When is work performance data collected?

PROJECT **Integration** MANAGEMENT

PROJECT **Integration** MANAGEMENT

Aɴsᴡᴇʀ

A progressive detailing of the project plan, providing detailed information about the work to be done in the current phase of the project and less information about the work to be done in later phases.

[Planning]

Aɴsᴡᴇʀ

During the direct and manage project work process

[Executing]

49

What are the four inputs to the direct and manage project work process?

50

The administrative closure procedure provides a step-by-step methodology that addresses three actions and activities. List them.

ANSWER

- Project management plan
- Approved change requests
- Enterprise environmental factors
- Organizational process assets

[Executing]

PROJECT **Integration** MANAGEMENT

ANSWER

- Satisfying completion and exit criteria for the phase or project
- Transferring the project's products, services, or results to the next phase or to production or operations
- Collecting phase or project records, auditing project successes or failures, gathering lessons learned, and archiving project information

[Closing]

PROJECT **Integration** MANAGEMENT

51

QUESTION

What are the five categories
of project management
processes?

52

QUESTION

What is a project management
office (PMO)?

ANSWER

- Initiating
- Planning
- Executing
- Monitoring and Controlling
- Closing

[Initiating, Planning, Executing, Monitoring and Controlling, and Closing]

ANSWER

An organization structure that standardizes project-related governance processes and facilitates the sharing of resources, methodologies, tools, and techniques

[Initiating, Planning, Executing, Monitoring and Controlling, and Closing]

53

QUESTION

At what level in the
organization should the
Project Sponsor be located?

PROJECT **Integration** MANAGEMENT

54

QUESTION

What are three project
documents that may be
updated as a result of the
monitor and control project
work process?

PROJECT **Integration** MANAGEMENT

ANSWER

At a level where the Project Sponsor is authorized to fund the project.

[Planning]

ANSWER

- Forecasts
- Performance reports
- Issue log

[Monitoring and Controlling]

55

What four tools and techniques are used to monitor and control project work?

56

What is the project statement of work and how is it used in the develop project charter process?

ANSWER

- Expert judgment
- Analytical techniques
- Project management information systems
- Meetings

[Monitoring and Controlling]

PROJECT **Integration** MANAGEMENT

ANSWER

A narrative description of the products, services, or results to be supplied by the project.

It is an input to the develop project charter process.

[Initiating]

PROJECT **Integration** MANAGEMENT

57

QUESTION

Name five subsidiary plans of
the project management plan.

58

QUESTION

What are three baselines that
form the project baseline?

ANSWER

Five of any of the following:

- Scope management
- Requirements management
- Schedule management
- Cost management
- Quality management
- Process improvement
- Human resource
- Communications management
- Risk management
- Procurement management
- Stakeholder management

[Planning]

ANSWER

- Scope baseline
- Schedule baseline
- Cost baseline

[Planning]

59 QUESTION

When is the project
management information
system first introduced?

60 QUESTION

Provide four examples of
items in the lessons learned
knowledge base.

Aɴsᴡᴇʀ

Direct and manage project work process

[Executing]

Aɴsᴡᴇʀ

- Project records and documents

- Project closure information and documents

- Results of previous project selection decisions and performance information

- Information from risk management activities

[Closing]

1

QUESTION

Define Project Scope
Management

2

QUESTION

What is the difference
between product scope and
project scope?

ANSWER

The processes required to ensure the project includes all the work required, and only the work required, to complete the project successfully.

[Planning and Monitoring and Controlling]

ANSWER

Product scope is the features and functions that characterize a product, service, or result, while project scope is the work performed to deliver a product, service, or result with the specified features and functions.

[Planning and Monitoring and Controlling]

3 QUESTION

What is a work package?

4 QUESTION

What is the purpose of define scope?

ANSWER

The work defined at the lowest level of the work breakdown structure for which cost and duration can be estimated and managed.

[Planning]

ANSWER

The process of developing a detailed description of the project and product.

[Planning]

5

QUESTION

What is control scope?

6

QUESTION

How is completion of product scope and project scope measured?

ANSWER

The process of monitoring the status of the project and product scope and managing changes to the scope baseline

[Monitoring and Controlling]

ANSWER

Project scope is measured against the project management plan, while product scope is measured against the product requirements.

[Planning and Monitoring and Controlling]

7 QUESTION

How is document analysis
used?

8 QUESTION

What is the purpose of the
collect requirements process?

ANSWER

As a tool and technique in collect requirements to elicit requirements by analyzing existing documents and identifying information that may be relevant to the requirements

[Planning]

ANSWER

To define and document stakeholders' needs to meet the project objectives.

[Planning]

9

QUESTION

Name any five of the 12 items that should be included in the WBS dictionary.

10

QUESTION

Name any five of the 11 tools and techniques in Collect Requirements

ANSWER

- Code of accounts identifier
- Description of work
- Assumptions and constraints
- Responsible organization
- Schedule milestones
- Associated schedule activities
- Resources required
- Cost estimates
- Quality requirements
- Acceptance criteria
- Technical requirements
- Agreement information

[Planning]

PROJECT **Scope** MANAGEMENT

ANSWER

- Interviews
- Focus groups
- Facilitated workshops
- Group creativity techniques
- Group decision-making techniques
- Questionnaires and surveys
- Observations
- Prototypes
- Benchmarking
- Context diagrams
- Document analysis

[Planning]

PROJECT **Scope** MANAGEMENT

11 QUESTION

What is the purpose of the WBS?

12 QUESTION

Define the create WBS process.

ANSWER

To show the total scope of work through a hierarchical decomposition to be done by the project team to accomplish the project objectives and create the required deliverables

[Planning]

ANSWER

Subdividing major project deliverables and project work into smaller, more manageable components.

[Planning]

13

QUESTION

What are the three items that
comprise the scope baseline?

14

QUESTION

When are user stories widely
used?

ANSWER

- Project scope statement
- WBS
- WBS dictionary

[Planning]

ANSWER

With agile methods as a way to collect requirements

[Planning]

15

QUESTION

What are the five inputs to the create WBS process?

16

QUESTION

What is the key benefit to the define scope process?

ANSWER

- Scope management plan
- Project scope statement
- Requirements documentation
- Enterprise environmental factors
- Organizational process assets

[Planning]

ANSWER

It defines the product, service, or result boundaries by defining which of the requirements will be included in or excluded from the project's scope.

[Planning]

17

QUESTION

How is the stakeholder
register used in collect
requirements?

18

QUESTION

What is the difference
between programs and
projects?

ANSWER

As an input to identify stakeholders that can provide information on detailed project and product requirements.

[Planning]

ANSWER

A program is a group of related projects, subprograms, and program activities managed in a coordinated way to obtain benefits not available from managing them individually. A project is a single element of work.

[Planning]

19

Describe two examples of
facilitated workshops.

20

How is the requirements
traceability matrix used in
validate scope?

A NSWER

- Joint application development or design (JAD) sessions: A collaboration technique first used in the software development industry to help users and the development team better, and more quickly, identify requirements.

- Quality Function Development (QFD): A quality technique often employed to identify critical characteristics for a new product.

[Planning]

A NSWER

As an input to link requirements to their origin and track them through the project life cycle to ensure they are included in the end product or service.

[Monitoring and Controlling]

21

QUESTION

List the two tools and
techniques for validate scope.

PROJECT **Scope** MANAGEMENT

22

QUESTION

What are two purposes of the
project scope statement?

PROJECT **Scope** MANAGEMENT

ANSWER

- Inspection
- Group decision-making techniques

[Monitoring and controlling]

ANSWER

- Describe the project's deliverables and the work required to complete them
- Provide a common understanding of the project scope across all stakeholder groups

[Planning]

23 QUESTION

When should the WBS
dictionary be updated?

24 QUESTION

What are five group creativity
techniques?

ANSWER

When approved change requests have an effect on the project scope.

[Monitoring and Controlling]

ANSWER

- Brainstorming
- Nominal group technique
- Idea/mind mapping
- Affinity diagram
- Multi-criteria decision analysis

[Planning]

25

QUESTION

Where should any funding
limitation be described?

26

QUESTION

What is the purpose of the
requirements management
plan?

In the project scope
statement.

[Planning]

To document and describe
how all requirements will
be analyzed, recorded, and
managed throughout the
project.

[Planning]

27

QUESTION

Why is the process of
preparing a WBS so important?

28

QUESTION

Describe three forms in which
the WBS structure can be
created.

ANSWER

To provide a structured vision
of what has to be delivered

[Planning]

ANSWER

- Using phases of the project
  life cycle at the second level
  with product and project
  deliverables at the third level

- Using major deliverables at
  the second level

- Incorporating
  subcomponents which
  may be developed by
  organizations outside of the
  project team; seller develops
  the supporting contact WBS
  as part of the contracted
  work

[Planning]

29

QUESTION

What are the five activities
involved in decomposition?

30

QUESTION

What are three organizational
process assets that can
influence the create WBS
process?

ANSWER

- Identify and analyze the deliverables and related work

- Structure and organize the WBS

- Decompose upper WBS levels into lower-level components

- Develop and assign identification codes to WBS components

- Verify that the degree of decomposition of work is necessary and sufficient

[Planning]

PROJECT **Scope** MANAGEMENT

ANSWER

- Policies, procedures, and WBS templates

- Project files from previous projects

- Lessons learned

[Planning]

PROJECT **Scope** MANAGEMENT

31 QUESTION

What is meant by progressive elaboration?

32 QUESTION

How are focus groups used in collect requirements?

ANSWER

Continuously improving and providing more detail to a plan as more information becomes available during project execution. The process produces a more accurate and complete plan as a result of such successive iterations.

[Planning]

ANSWER

To bring together prequalified stakeholders and subject matter experts to gain greater insight about their expectations and attitudes about the project's product, service, or end result.

[Planning]

33 QUESTION

What is the purpose of the
WBS dictionary?

34 QUESTION

What are five components
of the requirements
management plan?

ANSWER

To provide detailed deliverable, activity, and scheduling information about each component in the WBS

[Planning]

ANSWER

- How requirements will be planned, tracked, and reported
- Configuration management activities for requirements
- Requirements prioritization process
- Product metrics and the rationale for using them
- Traceability structure

[Planning]

35

QUESTION

What is the purpose of project exclusions? Where are they documented?

PROJECT **Scope** MANAGEMENT

36

QUESTION

What are six items included in the project scope statement?

PROJECT **Scope** MANAGEMENT

ANSWER

To identify what is explicitly outside the project's scope.

In the project scope statement.

[Planning]

ANSWER

- Product scope description
- Acceptance criteria
- Deliverables
- Project exclusions
- Constraints
- Assumptions

[Planning]

37

What is the purpose of
alternative generation in
scope definition?

38

How are group decision-
making techniques used in
validate scope?

ANSWER

To develop as many options as possible in order to identify different approaches to execute and perform the project work

[Planning]

ANSWER

As a tool and technique to reach a conclusion when the validation is performed by the project team and other stakeholders

[Monitoring and Controlling]

39 QUESTION

Define the validate scope process.

40 QUESTION

What are six examples of product analysis and where is it used?

Aɴsᴡᴇʀ

The process of formalizing acceptance of the completed project deliverables

[Monitoring and Controlling]

Aɴsᴡᴇʀ

- Product breakdown
- Systems analysis
- Requirements analysis
- Systems engineering
- Value engineering
- Value analysis

As a tool and technique in define scope.

[Planning]

41

What are four methods that can be used to reach a group decision, and where are they used?

PROJECT **Scope** MANAGEMENT

42

What is the difference between validate scope and control quality?

PROJECT **Scope** MANAGEMENT

ANSWER

- Unanimity
- Majority
- Plurality
- Dictatorship
- As a tool and technique in collect requirements

[Planning]

ANSWER

Validate scope is primarily concerned with acceptance of the deliverables; control quality is primarily concerned with the correctness of the deliverables and meeting the quality requirements of the deliverables.

[Monitoring and Controlling]

43 QUESTION

Why are observations helpful
in the collect requirements
process?

44 QUESTION

What is an example of a scope
model, and where is it used?

ANSWER

To view individuals in their actual work environment to see how they perform their jobs or tasks or otherwise execute processes. Observation is especially helpful if the people that use the product have difficulty or are reluctant to articulate their requirements.

[Planning]

ANSWER

The context diagram to visually depict the product scope by showing a business system and how people and other system (actors) interact with it.

Used as a tool and technique in collect requirements

[Planning]

45

QUESTION

What is meant by accepted
deliverables?

46

QUESTION

What are four terms that may
be used for inspections?

ANSWER

The acceptance criteria are formally signed off and approved by the customer or sponsor.

[Monitoring and Controlling]

ANSWER

- Reviews
- Product reviews
- Audits
- Walk-throughs

[Monitoring and Controlling]

47 QUESTION

How is work performance data used as an input to control scope?

48 QUESTION

Why is variance analysis an important tool and technique to control scope?

ANSWER

To show the number of change requests received; the number of requests accepted; or the number of deliverables completed

[Monitoring and Controlling]

ANSWER

To determine the cause and degree of difference between the baseline and actual performance

[Monitoring and Controlling]

49

QUESTION

What is the key benefit of validate scope?

50

QUESTION

What two project documents may be updated as an output of control scope?

ANSWER

To bring objectivity to the
acceptance process and
increase the chance of final
product, service, or result
acceptance by validating each
deliverable

[Monitoring and Controlling]

PROJECT **Scope** MANAGEMENT

ANSWER

- Requirements
  documentation

- Requirements traceability
  matrix

[Monitoring and Controlling]

PROJECT **Scope** MANAGEMENT

QUESTION

What are three examples of project constraints? Where are they documented?

QUESTION

What is the definition of uncontrolled changes?

PROJECT **Scope** MANAGEMENT

PROJECT **Scope** MANAGEMENT

Answer

- Predefined budget
- Imposed dates or scheduled milestones mandated by the customer or performing organization
- Contractual provisions (if the project is performed under contract) in the project scope statement.

[Planning]

Answer

Uncontrolled expansion to product or project scope without adjustments to time, cost, and resources

[Monitoring and Controlling]

53

What is another term for observations, and how is it done?

54

What are prototypes and how do they support progressive elaboration?

Job shadowing—usually done by an observer watching the user performing his or her job.

[Planning]

PROJECT **Scope** MANAGEMENT

Prototypes are working models of the end product.

Through interactive experimentation and feedback generation, the model is revised into the final product.

[Planning]

PROJECT **Scope** MANAGEMENT

55

How many levels are needed
in a WBS? Can the level of
detail vary?

56

What is the key benefit of
the plan scope management
process?

ANSWER

The number of WBS levels depends on the size and complexity of the project and the detail required to plan and manage it.

The level of detail may vary as the project evolves.

[Planning]

ANSWER

It provides guidance and direction on how scope will be managed throughout the process.

[Planning]

57

QUESTION

How should WBS components
be defined?

58

QUESTION

What are seven examples of
nonfunctional requirements?

ANSWER

An entry in the work breakdown structure that can be at any level

[Planning]

ANSWER

- Reliability
- Security
- Performance
- Level of service
- Safety
- Supportability
- Retention/purge

[Planning]

59

What are the four outputs from the validate scope process?

60

What is the Delphi Technique?

ANSWER

- Accepted deliverables
- Change requests
- Work performance information
- Project document updates

[Monitoring and Controlling]

ANSWER

An information gathering technique used to reach a consensus of experts on a subject; experts participate anonymously as a facilitator uses a questionnaire to solicit ideas.

Responses are only available to the facilitator, and consensus may be reached after several rounds in the process.

It helps to reduce bias in the data collected and keeps one person from dominating the process.

[Planning]

1 QUESTION

Project Time Management comprises which seven major processes?

2 QUESTION

Define total float.

ANSWER

- Plan Schedule Management
- Define Activities
- Sequence Activities
- Estimate Activity Resources
- Estimate Activity Durations
- Develop Schedule
- Control Schedule

[Planning and Monitoring and Controlling]

ANSWER

The amount of time that an activity can be delayed from its early start without delaying the project end date or violating a schedule constraint.

[Planning]

3 QUESTION

Why is the key benefit of the plan schedule management process?

4 QUESTION

How are performance reviews used in control schedule?

ANSWER

To provide guidance and direction on how the project schedule will be managed throughout the project

[Planning]

ANSWER

As a tool and technique to measure, compare, and analyze schedule performance such as actual start and finish dates, percent complete, and remaining duration for work in progress

[Monitoring and Controlling]

5

What is included in the define activities process?

6

What are the four project documents that may need to be updated as an output of sequence activities?

ANSWER

Identifying the specific actions to be performed in order to produce the project deliverables.

[Planning]

ANSWER

- Activity lists
- Activity attributes
- Milestone list
- Risk register

[Planning]

7

QUESTION

Describe the relationship
between the WBS and an
activity list.

8

QUESTION

What is another term used for
the precedence diagramming
method?

ANSWER

The activity list is developed from the identified work packages in the WBS.

[Planning]

ANSWER

Activity-on-node.

[Planning]

9

What is the result of a
critical path method (CPM)
calculation?

10

What is a planning package?

ANSWER

Early and late start and finish dates for each activity based on a specified network logic and a single deterministic duration estimate. The focus is on calculating *float* to determine which activities have the least scheduling flexibility.

[Planning]

ANSWER

A component below the control account but above the work package that identifies project work that does not have detailed schedule activities.

[Planning]

11

QUESTION

What is included in sequence activities?

12

QUESTION

What often happens as a result of resource leveling?

ANSWER

Identifying and documenting
relationships among the
project activities

[Planning]

ANSWER

The original critical path will
change.

[Planning]

13

Is elapsed time included when estimating required work periods?

14

Describe the four types of activity dependencies

Answer

Yes. For example, if "concrete curing" requires four days of elapsed time, it may require from two to four work periods, based on which day of the week it begins and whether weekend days are considered work periods.

[Planning]

Answer

- Mandatory dependencies – ones that are legally or contractually required or inherent in the nature of the work
- Discretionary dependencies – established based on knowledge of best practices or some unusual aspect of the project where a specific sequence is desired
- External dependencies – involve a relationship between project activities and non-project activities
- Internal dependencies – involve a precedence relationship between project activities and generally are within the team's control

[Planning]

15

What is the critical path?

16

List two tools and techniques used in plan schedule management.

ANSWER

The longest path through the network, which represents the shortest amount of time in which the project can be completed.

[Planning]

ANSWER

- Expert judgment
- Analytical techniques

[Planning]

17

QUESTION

How does resource leveling
work?

18

QUESTION

What are the nine sections
typically included in the
Schedule Management Plan?

Answer

Start and finish dates are adjusted based on resource constraints with the goal of balancing resource demand with available supply.

[Planning]

Answer

- Project schedule model development
- Level of accuracy
- Units of measure
- Organizational procedure links
- Project schedule model maintenance
- Control thresholds
- Rules of performance measurements
- Reporting formats
- Process descriptions

[Planning]

19

What is the purpose of estimate activity resources?

20

What does the following formula calculate?

[Optimistic + (4 × Most Likely) + Pessimistic] ÷ 6

ANSWER

Determine and identify
the types and quantities
of material, personnel,
equipment, or supplies
needed to complete each
activity.

[Planning]

ANSWER

An expected activity duration
using a weighted average
of three estimates. It is the
method used in the PERT
(Program Evaluation and
Review Technique).

[Planning]

21

QUESTION

List four resource categories found in a Resource Breakdown Structure.

22

QUESTION

What are three organizational process assets used in sequence activities?

ANSWER

- Labor
- Material
- Equipment
- Supplies

[Planning]

ANSWER

- Project files describing scheduling methodologies
- Activity related policies, procedures, and guidelines such as the scheduling methodology
- Templates

[Planning]

23

What is slack or float?
Is it different from lag?

24

What is a time-constrained
schedule?

Answer

The amount of time that a particular schedule activity can be delayed without delaying the project.

Yes. Lag is waiting time between activities in a network.

[Planning]

Answer

A project schedule that is fixed and cannot change. The unavailability of required resources for any activity is indicated by negative float.

[Planning]

25

Describe the four types of dependencies or precedence relationships in PDM. Of these, which one is the most commonly used dependency? Which one is the least commonly used?

26

What is the resource-constrained critical path called?

ANSWER

- Finish-to-start—The successor activity cannot start until the predecessor activity finishes

- Finish-to-finish—The successor activity cannot finish until the predecessor activity finishes

- Start-to-start—The successor activity cannot start until the predecessor activity starts

- Start-to-finish—The successor activity cannot finish until the predecessor activity starts

Finish-to-start

Start-to-finish

[Planning]

ANSWER

The critical chain.

[Planning]

27

What is crashing? What is its likely result?

28

What is the advantage of using a bar (Gantt) chart?

Aɴsᴡᴇʀ

A technique used to shorten the schedule for the least incremental cost by adding resources.

Crashing generally increases costs and results in increased risks.

[Planning]

Aɴsᴡᴇʀ

It is an effective progress-reporting tool, easily understood by most project stakeholders.

[Planning]

29 QUESTION

What is the schedule baseline?

30 QUESTION

What is the project buffer?

ANSWER

The approved version of a schedule model that can only be changed through formal change control procedures and is used as a basis for comparison to actual results.

[Planning]

ANSWER

An amount of time placed at the end of the critical chain to protect the target finish date from slippage along its path.

[Planning]

31

What are feeding buffers?

32

What are the disadvantages of using a bar (Gantt) chart?

ANSWER

The amount of time placed at each point that a chain of dependent tasks not on the critical chain feeds into the critical chain. By doing so, the critical chain is protected from slippage along the feeding chains.

[Planning]

ANSWER

- It is a weak planning tool.
- It does not show logical relationships between or among all project activities.

[Planning]

33

What is a resource breakdown structure? When is it prepared?

34

What is estimate activity durations? Who should do the estimating?

ANSWER

A hierarchical structure of the identified resources by resource category and resource type.

As an output from estimate activity resources.

[Planning]

ANSWER

Developing an approximation of the number of work periods needed to complete individual activities with estimated resources.

That person or persons most familiar with the work that needs to be performed.

[Planning]

35

What type of network logic diagram method was used to create the figure below?

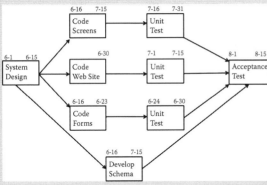

36

How is work performance data used in control schedule?

ANSWER

PDM

[Planning]

ANSWER

To provide project progress information such as which activities have started, finished, or are in progress.

[Monitoring and Controlling]

37

What is the difference between a milestone chart and a bar chart as it relates to develop schedule?

38

How does PERT differ from CPM?

ANSWER

The milestone chart identifies significant events such as a scheduled start of an activity, completion of a major deliverable, or approval to proceed by a higher authority. A bar chart graphically depicts an activity's start and end dates, including its estimated duration.

[Planning]

ANSWER

PERT uses the expected value (instead of the most likely estimate used in CPM) by using three time estimates per activity: pessimistic, most likely, and optimistic.

[Planning]

39

What are five tools and techniques used in estimate activity resources?

40

What PDM relationship is shown in the figure below?

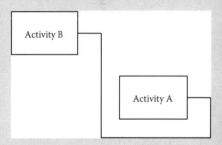

ANSWER

- Expert judgment
- Alternatives analysis
- Published estimating data
- Project management software
- Bottom-up estimating

[Planning]

ANSWER

Finish-to-finish.

[Planning]

41

What is fast-tracking? What is its likely result? What is a form of fast-tracking?

42

What is the focus of the critical chain network analysis technique?

ANSWER

Performing activities in parallel that normally would be performed in sequence.

It often results in rework and usually increases risk.

Concurrent engineering

[Planning]

ANSWER

To modify the project schedule to account for limited resources. Instead of managing the network paths' total float, it manages the remaining buffer durations against the remaining durations of task chains.

[Planning]

43

What is analogous estimating?

44

What are two performance measurements used to assess the magnitude of variation to the original schedule baseline?

ANSWER

Often called "top-down" estimating, it uses the actual duration of a previous, similar activity as the basis to estimate the duration of a future activity. It is a form of expert judgment. It is most reliable when the previous activities are similar in fact, not just in appearance, and when individuals preparing the estimate have the needed expertise.

[Planning]

ANSWER

- Schedule variance (SV)
- Schedule performance index (SPI)

[Monitoring and Controlling]

45

QUESTION

What is the purpose of the schedule management plan?

46

QUESTION

Describe the concept of a "range of results" in activity duration estimating.

ANSWER

It establishes the criteria and activities for developing, monitoring, and controlling the schedule.

[Planning]

ANSWER

A quantitative assessment of the likely number of work periods required to complete an activity. Example: 2 weeks plus or minus 2 days—the activity will take at least 8 days and no more than 12 days.

[Planning]

47

What is develop schedule?

48

When is the schedule baseline updated?

ANSWER

Analyzing activity sequences, durations, resource requirements, and schedule constraints to create the project schedule.

[Planning]

ANSWER

Immediately after approval is granted for change requests that impact project scope, activity resources, or activity duration estimates.

[Monitoring and Controlling]

49

What is the difference
between a project calendar
and a resource calendar?

50

What are three examples
of assumptions in estimate
activity durations?

ANSWER

A project calendar affects all resources. A resource calendar affects a specific resource or category of resources.

[Planning]

ANSWER

- Existing conditions
- Availability of information
- Length of the reporting periods

[Planning]

51

QUESTION

What is the difference between lead and lag?

52

QUESTION

How is simulation used in develop schedule? What technique is the most common?

ANSWER

Lead is a modification of a logical relationship that allows an acceleration of the successor task. Lag is a modification of a logical relationship that directs a delay in the successor task.

[Planning]

ANSWER

To calculate multiple project durations with different sets of activity assumptions.

Monte Carlo Analysis.

[Planning]

53

How is reserve analysis used in estimate activity durations? How is the schedule contingency reserve determined?

54

How is the scope baseline used in define activities?

ANSWER

To incorporate additional time into the project schedule as a reflection of schedule risk.

Contingency reserve can be a percentage of the estimated activity or project duration, a fixed and agreed-to number of work periods, or can be developed by using other quantitative methods.

[Planning]

ANSWER

As an input to the process.

The WBS, deliverables, constraints, and assumptions documented in the scope baseline are considered as activities are defined.

[Planning]

55

What estimating problem is addressed through the critical chain method?

56

What should always be included as part of the supporting detail for the project schedule?

ANSWER

The tendency by those persons responsible for executing activities to be too conservative in their duration estimating, resulting in extended project durations and overestimation of required production resources such as staff specialists or unique equipment.

[Planning]

ANSWER

- Detailed resource requirements by time period

- Alternative schedules, such as best and worst case, not-resource-leveled, or resource-leveled dates

- How contingency reserves are scheduled

[Planning]

57

QUESTION

How are performance reviews used in Control Schedule when the critical chain method is being used?

58

QUESTION

What is an example of an internal dependency?

ANSWER

To help determine schedule status one can compare the amount of buffer remaining to the amount of buffer needed to protect the delivery date. Corrective action may be appropriate depending on the difference between the buffer needed and the buffer remaining.

[Monitoring and Controlling]

ANSWER

If the project team cannot test a machine until it is developed

[Planning]

59

How is what-if scenario analysis used in develop schedule?

60

What are three examples of organizational process assets that may be updated as a result of control schedule?

ANSWER

To test whether the project's schedule is realistic under adverse conditions, and to prepare back-up plans to mitigate the impact of unexpected situations.

[Planning]

ANSWER

- Causes of variances
- Corrective actions chosen and the reasons
- Other types of lessons learned from project schedule control

[Monitoring and Controlling]

1

QUESTION

Define Project Cost Management.

2

QUESTION

What is the key benefit of Plan Cost Management?

ANSWER

Processes involved in planning, estimating, budgeting, and controlling costs so the project can be completed within the approved budget

[Planning and Monitoring and Controlling]

ANSWER

To provide guidance and direction on how the project costs will be managed throughout the project

[Planning]

3

QUESTION

Identify four organizational process assets used in estimate costs.

4

QUESTION

List three general financial management techniques when predictions and analyses are included in Project Cost Management.

ANSWER

- Cost estimating policies
- Cost estimating templates
- Historical information
- Lessons learned

[Planning]

ANSWER

- Return on investment
- Discounted cash flow
- Investment payback analysis

[Planning]

5

QUESTION

What is the BAC? What
question does it answer?

6

QUESTION

What is the payback period?

ANSWER

Budget at completion. It is the total Planned Value for the project and constitutes the cost performance baseline.

What *is* the total job *supposed* to cost?

[Planning and Monitoring and Controlling]

ANSWER

The number of time periods up to the point at which cumulative revenues exceed cumulative costs and, therefore, the project has "turned a profit."

[Planning]

7 QUESTION

How is the WBS used in
determine budget?

8 QUESTION

What is the internal rate of
return (IRR)?

ANSWER

As an input the process.

The WBS provides the relationships among all project deliverables and their various components.

[Planning]

ANSWER

The percentage rate that makes the present value of costs equal to the present value of benefits.

[Planning]

9

What is the AC? What question does it answer?

10

What is PV? What question does it answer?

ANSWER

AC means actual costs or the actual cost incurred in accomplishing work on a WBS activity component or on the total project.

How much did the completed work cost?

[Monitoring and Controlling]

ANSWER

Planned value.

It is the authorized budget assigned to the work scheduled to be accomplished for an activity or WBS component.

How much work should be done?

[Monitoring and Controlling]

11

QUESTION

List two enterprise environmental factors that can be used in estimate costs.

12

QUESTION

When do activity duration estimates affect project cost estimates?

ANSWER

- Marketplace conditions
- Published commercial information

[Planning]

ANSWER

- When "cost of financing" project activities is a reality
- When costs are based on the number of units of time estimated
- When costs are sensitive to time such as union labor rates increasing on a certain date

[Planning]

13

QUESTION

What is the EAC? What
question does it answer?

14

QUESTION

What is ETC? What question
does it answer?

ANSWER

Estimate at completion—
the expected total cost of
a schedule activity, a WBS
component, or the project at
completion.

What do we *now* expect the
total job to cost?

[Monitoring and Controlling]

ANSWER

Estimate to complete the
expected cost to complete
a schedule activity, a WBS
component, or the project.

What do we need to spend to
complete the remaining work?

[Monitoring and Controlling]

15 QUESTION

What is EV? What question does it answer?

16 QUESTION

What is the "50-50" rule of progress reporting? What assumption underlies this rule?

ANSWER

Earned value—the value of the work performed expressed in terms of the approved budget assigned to that work for a schedule activity or WBS component.

How much work is done?

[Monitoring and Controlling]

ANSWER

When beginning a task, charge 50 percent of its PV to its account; when the task is completed, charge the remaining 50 percent to its account.

All tasks generally are of the same size.

[Monitoring and Controlling]

17

QUESTION

Identify ten estimate cost tools and techniques

18

QUESTION

How is schedule variance (SV) calculated?

ANSWER

- Expert judgment
- Analogous or top-down estimating
- Parametric estimating
- Bottom-up estimating
- Three-point estimating
- Reserve analysis
- Cost of quality
- Project management estimating software
- Vendor bid analysis
- Group decision-making techniques

[Planning]

ANSWER

$SV = EV - PV$

[Monitoring and Controlling]

19

If you are using the payback period to compare projects A and B, and project A has the shortest payback period, what does this mean?

20

What are direct costs?

ANSWER

Project A becomes profitable
more quickly.

[Planning]

ANSWER

Costs incurred directly by a
specific project.

[Planning]

21

Analogous estimating is most reliable when two conditions are met. Describe them.

22

What is parametric cost estimating, and how can it provide higher levels of accuracy?

ANSWER

- When the previous projects have similar characteristics as the current project

- When the person(s) preparing the estimates has the required expertise

[Planning]

ANSWER

It uses a statistical relationship between relevant historical data and other variables to calculate a cost estimate for the project work. Higher levels of accuracy are achieved based on the sophistication of the data in the model.

[Planning]

23

Describe bottom-up estimating.

24

What are management reserves? Are they in the cost baseline?

Estimating and then summarizing or rolling up the cost of individual activities or work packages to get a project total.

[Planning]

An amount of the total budget withheld for management purposes for unforeseen work or for "unknown unknowns"

They are not part of the cost baseline but are part of the overall project budget and funding requirements.

[Planning]

25

What are variable costs?

26

List the five recommended types of supporting detail for activity cost estimates.

ANSWER

Costs that rise or fall directly with the size of the project.

[Planning]

ANSWER

- The basis for the estimate
- The assumptions made
- The constraints
- An indication of the range of results (e.g., $5,000 \pm 10\%$)
- An indication of the confidence level of the final estimate

[Planning]

27

QUESTION

How is the cost performance index (CPI) calculated?

28

QUESTION

List nine items to be part of the cost management plan.

$CPI = EV / AC$

[Monitoring and Controlling]

- Units of measure
- Level of precision
- Level of accuracy
- Organizational procedures link
- Control thresholds
- Rules of performance measurement
- Reporting formats
- Process descriptions
- Additional details

[Planning]

29

What does the cost
management plan describe?

30

What is the variance at
completion (VAC)? How is it
calculated?

ANSWER

How the project costs will be planned, structured, and controlled

[Planning]

ANSWER

The difference between the total amount the job was supposed to cost (BAC), and the amount the job is now expected to cost (EAC).

$VAC = BAC - EAC$

[Monitoring and Controlling]

31

QUESTION

What is the status of this project?

32

QUESTION

What is determine budget?

ANSWER

Over budget (AC is higher than PV) and behind schedule (EV is less than PV).

[Planning]

ANSWER

Aggregating the estimated costs of all individual activities or work packages in the WBS to establish an authorized cost baseline.

[Planning]

33

What are fixed costs? Are they recurring or nonrecurring costs?

34

What is a rough order-of-magnitude estimate? When is it performed? What is its accuracy range?

Costs that do not change based on the number of units.

Nonrecurring.

[Planning]

An approximation without detailed data.

Often done early in a project when you need a "ballpark estimate."

±50 percent.

[Planning]

35

What are the nine inputs to determine budget?

36

What are indirect costs?

ANSWER

- Cost management plan
- Scope baseline
- Activity cost estimates
- Basis of estimates
- Project schedule
- Resource calendars
- Risk register
- Agreements
- Organizational process assets

[Planning]

ANSWER

Costs that are part of the overall organization's cost of doing business and are shared among (allocated to) all the projects that are under way.

[Planning]

37

QUESTION

How is a definitive estimate prepared? Give an example of one. What is its accuracy range?

38

QUESTION

What is the cost baseline?

ANSWER

From well-defined, detailed data.

A bottom-up estimate.

–5 to +10 percent.

[Planning]

ANSWER

Authorized version of the time-phased budget, excluding management reserve, which can only be changed through formal change control processes and is used for comparison to actual results.

[Planning]

39

How is the SPI calculated?

What does it tell us?

What does an SPI less than 1.0 mean?

40

How is CPI calculated?

What does it tell us?

What does a CPI greater than 1.0 mean?

ANSWER

SPI = EV / PV

As a progress measure it tells us how much work has been achieved compared to the plan.

Less work was completed than planned.

[Monitoring and Controlling]

ANSWER

CPI = EV / AC

Measure of work value completed compared to actual cost of completed work.

Cost under-run.

[Monitoring and Controlling]

41

What are the four organizational process assets used for the plan cost management process?

42

What are the four inputs to the control costs process?

ANSWER

- Financial control procedures
- Historical information and lessons learned knowledge bases
- Financial databases
- Existing cost estimating and budgeting-related policies, procedures, and guidelines

[Planning]

ANSWER

- Project management plan
- Project funding requirements
- Work performance data
- Organizational process assets

[Monitoring and Controlling]

43 QUESTION

What is present value (PrV)?
How is it calculated?

44 QUESTION

When is rebaselining required?

Answer

The value today of future cash flows based on the concept that payment today is worth more than payment tomorrow.

$$PrV = \frac{M_t}{(1+r)^t}$$

M_t = amount of payment t years from now
r   = interest rate (or discount rate)
t   = time period

[Planning]

Answer

When the cost variances are so severe that a complete revision of the project budget is needed to provide a realistic measure of performance.

[Monitoring and Controlling]

45

How are group decision-
making techniques used in
estimating costs?

46

List four ways to calculate EAC.

ANSWER

To improve accuracy and commitment to the emerging estimates

[Planning]

ANSWER

- EAC = BAC/CPI

- EAC = AC + BAC – EV

- EAC = AC + BottomupETC

- EAC = AC + [(BAC – EV)/(CPI × SP

[Monitoring and Controlling]

47 QUESTION

What is the basis for a budget estimate?

When is it used?

What is its accuracy range?

48 QUESTION

What is the to-complete-performance index (TCPI)?

What does it tell us?

Is it based on BAC or EAC?

ANSWER

This type of estimate is based on data that are more detailed than those used for an order-of-magnitude estimate.

Used to establish initial funding and to gain project approval.

–10 to +25 percent.

[Planning]

ANSWER

- The ratio of the "remaining work" to the "funds remaining."

- Tells us the cost performance that must be achieved on the remaining work to meet a specific goal.

- It can be based on either BAC or EAC. If BAC is viable, equation is:

$BAC = (BAC - EV) / (BAC - AC)$

If BAC is not viable, equation is:

$EAC = (BAC - EV) / (EAC - AC)$

[Monitoring and Controlling]

49 QUESTION

Why is the management reserve added to cost baseline?

50 QUESTION

What is value analysis?

ANSWER

To produce the project budget.

[Planning]

ANSWER

A cost-reduction tool that involves careful analysis of a design or item to identify all the functions and the cost of each one. It considers whether the function is necessary and whether it can be provided at a lower cost without degrading performance or quality.

[Planning]

51

How is cost variance (CV) calculated?

52

Which sections of the project management plan are used in the control cost process?

ANSWER

CV = EV – AC

[Monitoring and Controlling]

ANSWER

Cost management plan

Cost baseline

[Monitoring and Controlling]

53

When should EAC be
calculated by: BAC/CPI?

54

When should EAC be
calculated by: AC + BAC − EV?

ANSWER

If the CPI is expected to be the same for the remainder of the project.

[Monitoring and Controlling]

ANSWER

If future work will be accomplished at the planned rate

[Monitoring and Controlling]

55

When should EAC
be calculated by:
AC + Bottomup ETC?

56

What is the difference
between variance analysis and
trend analysis in project cost
management?

If the initial plan is no longer valid

[Monitoring and Controlling]

PROJECT **Cost** MANAGEMENT

ANSWER

Variance analysis is a technique that compares actuals to planned data to determine degree of variation to cost baseline.

Trend analysis looks at data over a specified time period to see if performance is improving or deteriorating.

[Monitoring and Controlling]

PROJECT **Cost** MANAGEMENT

57 QUESTION

What drives the cost and accuracy of bottom-up estimating?

58 QUESTION

Define control costs.

ANSWER

The size and complexity of the individual activity or work package.

[Planning]

ANSWER

The process of monitoring project status to update the project costs and manage changes to the cost baseline

[Monitoring and Controlling]

59

QUESTION

What does the benefit–cost ratio not tell you?

60

QUESTION

When should EAC be calculated by:

$EAC = AC + [(BAC – EV)/(CPI \times SPI)]$?

ANSWER

The absolute value of the benefit or cost.

[Planning]

ANSWER

If both the CPI and SPI influence the remaining work

[Monitoring and Controlling]

PROJECT **Quality** MANAGEMENT

1

What is Project Quality Management?

2

Three standard deviations on either side of the mean of a normal distribution will contain approximately what percentage of the total population?

Aɴꜱᴡᴇʀ

The performing organization's processes and activities that determine quality policies, objectives, and responsibilities, which ensure that the project will satisfy the needs for which it was undertaken.

[Planning, Executing, and Monitoring and Controlling]

Aɴꜱᴡᴇʀ

99.7%

[Monitoring and Controlling]

3

What are control limits as used
in control charts?

4

What is the difference
between grade and quality?

ANSWER

Control limits (upper and lower) define the natural variation of a process. Points within the limits generally indicate normal and expected variation. Points outside the limits generally indicate that something has occurred that requires special attention because it is outside the natural variation in the process.

[Planning, Executing, and Monitoring and Controlling]

ANSWER

Grade is a category assigned to a product or service whose functional use is similar but whose technical characteristics differ. Quality is the degree to which a set of characteristics fulfills requirements. Low quality always is a problem; low grade may not be.

[Planning and Monitoring and Controlling]

5 QUESTION

What tool is used to present the information below?

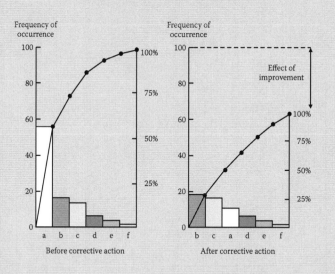

6 QUESTION

Define *kaizen*.

ANSWER

A Pareto diagram (or chart).

[Planning, Executing, and Monitoring and Controlling]

ANSWER

It is the Japanese term for continuous improvement.

[Executing]

7

QUESTION

How does the temporary
nature of a project affect
quality management?

8

QUESTION

What is gold-plating?
Describe its value.

ANSWER

Investments in product quality improvement, such as defect prevention and appraisal, must be borne by the performing organization because the project may not last long enough to realize rewards.

[Executing]

ANSWER

Giving the customer more than what was required.

It has no value; exceeding specified requirements is a waste of time and money with no value added to the project. The customer should expect and receive exactly what was specified—no more, no less.

[Planning]

9

What is the purpose of a control chart?

10

What is plan quality management?

ANSWER

To determine whether a process is stable or has predictable performance.

[Planning, Executing, and Monitoring and Controlling]

ANSWER

Identifying project quality requirements and/or standards for project deliverables and documenting how the project will demonstrate compliance with quality requirements.

[Planning]

11

QUESTION

Is perform quality assurance a managerial or technical function?

12

QUESTION

What should the project manager do if the performing organization does not have a quality policy?

ANSWER

Managerial.

[Executing]

ANSWER

Work with the project management team to develop a quality policy for the project and ensure that project stakeholders are aware of the quality policy.

[Planning]

13

What is the purpose of a
flowchart?

PROJECT **Quality** MANAGEMENT

14

What is the effect of sample
size on the standard deviation?

PROJECT **Quality** MANAGEMENT

ANSWER

To graphically depict the relationships among steps in a process (including activities, decisions points, and the order of processing). In planning, flowcharting helps the project team anticipate quality problems. In monitoring and controlling, it can pinpoint a failing process step(s) and help identify potential process improvement opportunities.

[Planning, Executing, and Monitoring and Controlling]

ANSWER

Whenever sample size increases, the standard deviation decreases.

[Monitoring and Controlling]

15

Why is the project scope statement a key input to plan quality management?

16

What is the Rule of Seven?

ANSWER

It documents the product description, major project deliverables, and acceptance criteria. It often contains details of technical issues and other concerns that can affect quality planning.

[Planning]

ANSWER

It means that there are seven consecutive points (measurements) outside the upper or lower control limits in a control chart, indicating a process that is out of control.

[Planning, Executing, and Monitoring and Controlling]

17

QUESTION

On a scatter diagram, what is the significance of the closer the points are to a diagonal line?

18

QUESTION

What is rework?

ANSWER

The more closely the two variables measured are related.

[Planning, Executing, and Monitoring and Controlling]

ANSWER

Action taken to ensure that a defective or nonconforming item complies with requirements or specifications.

[Monitoring and Controlling]

19

What is benchmarking?

20

Define quality.

ANSWER

Compares actual or planned practices to those of comparable organizations to identify best practices, generate ideas for improvement, and provide a basis for measuring performance

[Planning, Executing, and Monitoring and Controlling]

PROJECT **Quality** MANAGEMENT

ANSWER

"The degree to which a set of inherent characteristics fulfils requirements." (ISO 9000)

[Planning, Executing, and Monitoring and Controlling]

PROJECT **Quality** MANAGEMENT

21

Name the two major categories of the cost of nonconformance. Give at least two examples of each.

22

List the seven basic tools of quality.

ANSWER

Internal Failure Costs
- Scrap
- Rework

External Failure Costs
- Liabilities
- Warranty work
- Lost business

[Planning]

ANSWER

- Cause-and-effect diagrams
- Flowcharts
- Checksheets
- Pareto diagrams
- Histograms
- Control charts
- Scatter diagrams

[Planning, Executing, and Monitoring and Controlling]

QUESTION

What is meant by cost of quality?

QUESTION

How are cause-and-effect diagrams used in Project Quality Management?

PROJECT **Quality** MANAGEMENT

PROJECT **Quality** MANAGEMENT

Aɴsᴡᴇʀ

The total cost of all efforts related to quality throughout the product life cycle, including the cost of conformance and the cost of nonconformance.

[Planning]

Aɴsᴡᴇʀ

As a tool and technique in all three processes to describe a problem as a gap to be closed or an objective to be achieved. The causes are found by looking at the problem statement and asking why until actionable root causes are identified, or reasonable possibilities on the diagram have been exhausted.

[Planning, Executing, and Monitoring and Controlling]

25

What is the voice of the customer (VOC)?

26

What is meant by design of experiments? Where is it most frequently applied?

ANSWER

Answer: A planning technique used to provide products, services, or results reflecting customer requirements by translating appropriate ones into technical requirements for each phase of the project or product.

[Planning]

ANSWER

Analytical technique used to help identify the variables that have the most influence on the outcome of a process or procedure. As an example, it is used during the plan quality management process to determine the number and type of tests and their impact on the cost of quality.

[Planning]

27

What is the difference between an attribute and a variable?

28

What is the principal purpose of the quality management plan?

ANSWER

An attribute is a quality characteristic that is classified as either conforming or nonconforming to specifications or requirements.

A variable is a quality characteristic that is rated on a continuous scale that measures the degree of conformity to the specifications or requirements.

[Monitoring and Controlling]

ANSWER

To describe how the organization's quality policies will be implemented and how the project management team plans to meet the quality requirements set for the project

[Planning]

29

What are quality checksheets?
How are they used?

30

The upper and lower control
limits are set at what level of
standard deviation?

PROJECT **Quality** MANAGEMENT

PROJECT **Quality** MANAGEMENT

ANSWER

As a checklist in gathering data to organize facts in a way that will facilitate effective collection of useful data about a quality problem.

They are one of the seven basic quality tools and are used throughout project quality management.

[Planning, Executing, and Monitoring and Controlling]

ANSWER

± 3 s

[Planning, Executing, and Monitoring and Controlling]

Define perform quality
assurance.

What is a quality audit?
What are its five objectives?

ANSWER

The process of auditing quality requirements and results from quality control measurements to ensure that appropriate standards and operational definitions are used.

[Executing]

ANSWER

Structured review of quality management activities to ensure project activities are in compliance with all policies, processes, and procedures.

- Identifies good/best practices
- Identifies gaps/shortcomings
- Shares best practices with other projects across the organization
- Offers process improvement assistance
- Ensures updating of the organization's lessons learned repository

[Executing]

PROJECT **Quality** MANAGEMENT

PROJECT **Quality** MANAGEMENT

33

Who is ultimately responsible for providing the resources needed to ensure quality in any given project?

34

What is control quality?

ANSWER

Management – usually the project sponsor.

[Planning, Executing, and Managing and Controlling]

ANSWER

Monitoring and recording results of executing the quality activities to assess performance and recommend necessary changes.

[Monitoring and Controlling]

35 QUESTION

Give seven examples of costs
of conformance.

36 QUESTION

In which project phase is
control quality performed?

Aɴsᴡᴇʀ

- Training
- Document processes
- Equipment
- Time to do it right
- Testing
- Destructive testing loss
- Inspections

[Planning]

Aɴsᴡᴇʀ

Control quality is performed throughout all phases of the project life cycle.

[Monitoring and Controlling]

37

What is a Pareto diagram?
How is it used in control quality?

38

When is a process "in control"?
"Out of control"?

ANSWER

A special form of vertical bar chart used to identify the vital few sources responsible for causing cost of a problem's effect; organized into categories to measure either frequencies or consequences.

It puts issues into an easily understood framework in which rank ordering is used to focus corrective actions.

[Monitoring and Controlling]

ANSWER

In control when the process is within acceptable limits – within the upper and lower limits on a control chart

Out of control if a data point exceeds a control limit; seven consecutive plot points are above the mean; or seven consecutive plot points are below the mean

[Planning, Executing, and Monitoring and Controlling]

39

QUESTION

In what process should sample frequency and sizes be determined?

40

QUESTION

What process improvement model is used as the basis for quality improvement as defined by Shewhart?

ANSWER

Plan quality management process.

[Planning]

ANSWER

Plan-do-check-act cycle.

[Planning]

41

QUESTION

What is the difference
between a special cause and a
random cause?

42

QUESTION

What are the updates to
organizational process assets
as a result of perform quality
control?

ANSWER

A special cause is an unusual event and indicates a process is "out of control"; a random cause is normal process variation, and indicates a process is "in control."

[Monitoring and Controlling]

ANSWER

- Completed checklists
- Lessons learned documentation

[Monitoring and Controlling]

43 QUESTION

What is the difference
between specification limits
and control limits?

44 QUESTION

What is inspection?

ANSWER

Specification limits establish the range of acceptable results. Control limits are a threshold that can indicate if the process is out of control.

A process could be in control in which the results fall within the computed control limits, but are outside the acceptable specification limits, which results in an unacceptable outcome.

[Planning, Executing, and Monitoring and Controlling]

PROJECT **Quality** MANAGEMENT

ANSWER

Examining or measuring whether an activity, component, product, result, or service conforms to specified requirements.

[Monitoring and Controlling]

PROJECT **Quality** MANAGEMENT

45

What is root cause analysis?

46

What is the underlying concept of a Pareto diagram?

ANSWER

An analytical technique used to determine the basic underlying reason that causes variance, defect, or a risk. A root cause may underlie more than one variance, defect, or risk.

[Planning, Executing, and Monitoring and Controlling]

ANSWER

A relatively large majority of the problems or defects (80 percent) are the result of a small number of causes (20 percent). Also called the "80/20" rule.

[Planning, Executing, and Monitoring and Controlling]

47 QUESTION

What is the purpose of approved changed requests review in perform quality control?

48 QUESTION

What are verified deliverables in perform quality control?

ANSWER

It is a tool and technique to ensure all approved change requests were implemented.

[Monitoring and Controlling]

ANSWER

A goal of the process is to determine the correctness of deliverables. The result of performing this process is verified deliverables, which are an input to Validate Scope.

[Monitoring and Controlling]

49

QUESTION

Who developed the tool represented by the following figure? What are three names for it?

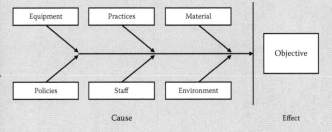

50

QUESTION

What do the upper control limit and lower control limit represent on this figure?

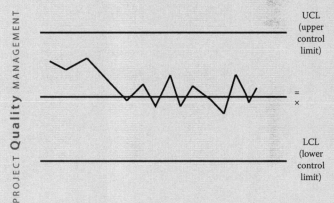

Ishikawa

- Cause-and-effect diagrams
- Ishikawa diagrams
- Fishbone diagrams

[Planning, Executing, and Monitoring and Controlling]

3σ from the mean.

[Planning, Executing, and Monitoring and Controlling]

51

QUESTION

Modern quality management emphasizes some of the same basic principles of project management. Provide five examples.

52

QUESTION

What are ten additional quality planning tools that often are used to better define quality requirements and plan effective quality management activities?

ANSWER

- Customer satisfaction
- Prevention over inspection
- Continuous improvement
- Management responsibility
- Cost of quality

[Planning, Executing, and Monitoring and Controlling]

ANSWER

- Brainstorming
- Force field analysis
- Nominal group technique
- Affinity diagrams
- Process decision program charts
- Interrelationship diagraphs
- Tree diagrams
- Prioritization matrices
- Activity network diagrams
- Matrix diagrams

[Planning and Executing]

53

What is a quality metric?
Provide seven examples.

54

What is the primary benefit of
meeting quality requirements?
What is the primary cost?

ANSWER

An operational definition that describes in specific terms a project or product attribute and how the control quality process will measure it.

- On-time performance
- Budget control
- Defect frequency
- Failure rate
- Availability
- Reliability
- Test coverage

[Planning, Monitoring and Controlling]

ANSWER

The primary benefit is less rework, which translates into higher productivity, lower costs, and increased stakeholder satisfaction.

The primary cost is the expense associated with project quality management activities.

[Planning]

55

In what processes are quality checksheets used?

56

How are quality control measurements used in perform quality assurance?

ANSWER

- Plan Quality Management
- Perform Quality Assurance
- Perform Quality Control

[Planning, Executing, and Managing and Controlling]

ANSWER

To analyze and evaluate the project's processes against the standards of the performing organization or the specified requirements

[Executing]

57

QUESTION

Perform quality assurance is performed in what phase?

58

QUESTION

What is process analysis? When is it used?

ANSWER

Throughout all phases of the project life cycle.

[Executing]

ANSWER

Examination of the individual steps in a process to identify needed improvements. The analysis examines problems experienced, constraints experienced, and non-value-added activities identified during process operation. It includes root cause analysis.

As a tool and technique in perform quality assurance.

[Executing]

59

What are four areas to consider
in the process improvement
plan?

60

If a project is experiencing
quality problems, should the
project manager devote more
project resources to inspection
or to prevention?

ANSWER

- Process boundaries
- Process configuration
- Process metrics
- Targets for improved performance

[Executing]

ANSWER

Prevention, because preventing a problem is far less costly, in the long run, than fixing one, especially if the customer finds the problem.

[Planning]

1
QUESTION

Project Human Resource
Management comprises which
four processes?

2
QUESTION

What are two key items in
managing and leading the
project team?

ANSWER

- Plan Human Resource Management
- Acquire Project Team
- Develop Project Team
- Manage Project Team

[Planning and Executing]

ANSWER

- Influencing the team
- Professional and ethical behavior

[Planning and Executing]

3

QUESTION

One way to resolve conflict is through collaboration. Why?

4

QUESTION

Describe the difference between a role and a responsibility.

ANSWER

It incorporates multiple viewpoints and insight from different perspectives, requires a cooperative attitude and open dialogue, and typically leads to consensus and commitment. It may be called problem solve.

[Executing]

ANSWER

A role is a defined function to be performed.

A responsibility is an assignment that can be delegated and that the assigned person then has a duty to perform the assignment.

[Planning]

5

QUESTION

From the project manager's perspective, give four disadvantages of the functional organization.

6

QUESTION

What are the five enterprise environmental factors that are inputs in the plan human resource management process?

ANSWER

- No formal authority over project resources

- Reliance on the informal power structure and his or her own interpersonal skills to obtain resource commitments from functional managers

- Little, if any, control over the budget

- The project manager works on the project part time

[Planning]

ANSWER

- Organizational culture and structure

- Existing human resources

- Geographic dispersion of team members

- Personnel administration policies

- Marketplace conditions

[Planning]

7 QUESTION

Name four sources of conflict
in projects.

8 QUESTION

Describe the purpose
of ground rules. Who is
responsible for enforcing them
once they are established?

ANSWER

- Scarce resources
- Scheduling
- Priorities
- Personal work styles

[Executing]

ANSWER

To establish clear expectations regarding acceptable team member behavior. They can decrease misunderstandings and increase productivity.

The project team is responsible for enforcing them.

[Executing]

9

What is McGregor's Theory X? What is management's role in this approach?

10

What are two useful templates in the plan human resource management process?

ANSWER

The traditional approach to managing workers: Workers are seen as inherently self-centered, lazy, and lacking in ambition; a top-down view of how people should be managed.

Managers organize the elements of the productive enterprise in the interest of economic ends.

[Planning]

ANSWER

- Project organization charts
- Position descriptions

[Planning]

11

QUESTION

Who is responsible for conducting formal or informal assessments of the project team's performance?

12

QUESTION

What is the difference between a responsibility assignment matrix and a staffing management plan?

ANSWER

Project management team.

[Planning and Executing]

ANSWER

A responsibility assignment matrix graphically depicts which work packages or activities are completed by which project team member. A staffing management plan describes when and how human resource requirements will be met.

[Planning]

13

QUESTION

Provide seven ways expert judgment can be used in developing the human resource management plan.

14

QUESTION

What are four indicators of a team's effectiveness?

Aɴsᴡᴇʀ

- List preliminary requirements for required skills

- Assess roles needed for the project

- Determine the preliminary effort level and number of resources needed to meet objectives

- Determine reporting relationships

- Provide guidelines on leadtime for staffing

- Identify risks associated with staff acquisition, retention, and release plans

- Identify and recommend programs for compliance with any applicable government and union contracts

[Planning]

Aɴsᴡᴇʀ

- Individual skills improvement so team members can perform assignments more effectively

- Improvements in competencies that raise overall team performance

- Reduced staff turnover rate

- Increased team cohesiveness

[Executing]

15

In what stage of team development do the members of the project team begin to trust each other?

16

Does the matrix form of project organization facilitate or complicate project team development? Why?

ANSWER

Norming.

[Executing]

ANSWER

It complicates project team development.

Because team members are accountable to both a functional manager and a project manager.

[Planning]

17

QUESTION

What is the purpose of a resource breakdown structure (RBS)?

18

QUESTION

Why is it important to consider cultural differences when determining recognition and rewards?

ANSWER

To break down the project by resource categories and types of resources.

It is used in resource leveling to develop resource-limited schedules, and to identify and analyze project human resource assignments.

[Planning]

ANSWER

The project manager would not want to give team-based awards in a culture that values individual achievement. This would be contrary to cultural traditions.

[Executing]

19

QUESTION

How would you describe forcing as a means of managing conflict?

20

QUESTION

What does this table represent?

PHASE \\ PERSON	X	Z	G	F	S	T
Concept	A	R	I	C	R	
Requirements	C		I	C	I	A
Design	A		R	A		
Development		R	C	A		C
Implementation			C	I	R	I

R = Responsible   A = Accountable
C = Consult   I = Inform

ANSWER

Exerting one's view at the potential expense of another party, which establishes a win-lose situation.

[Executing]

ANSWER

A RACI (Responsible, Accountable, Consult, Inform) chart; a type of responsibility assignment matrix (RAM).

[Planning]

21

QUESTION

What is McGregor's Theory Y?
What is management's role in
this approach?

22

QUESTION

Name three major forms
of project organizational
structure.

Answer

Workers are not by nature resistant to organizational needs; they are willing and eager to accept responsibilities and are concerned with self-growth and self-fulfillment.

Managers should try to create an environment where workers can achieve their own goals while directing efforts toward organizational objectives.

[Planning]

Answer

- Functional
- Matrix
- Projectized

[Planning]

23 QUESTION

What does this figure represent?

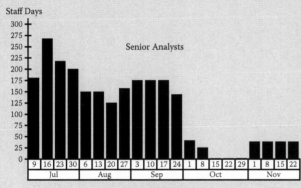

Staff Days

Senior Analysts

Resource Usage

24 QUESTION

When should the project expeditor form of organization be used?

ANSWER

A resource histogram.

[Planning]

ANSWER

When a project's cost and importance are relatively low.

[Planning]

25

What are three components of the human resource plan?

26

What area of project management becomes increasingly important when managing a virtual team?

ANSWER

- Roles and responsibilities
- Project organization charts
- Staffing management plan

[Planning]

ANSWER

Communications planning.

[Planning]

27

QUESTION

What is the expectancy
theory?

28

QUESTION

How does the project
manager's authority differ
in various organizational
structures?

ANSWER

It holds that people tend to be highly productive and motivated if they believe their efforts will lead to successful results, and they will be rewarded for their success.

[Planning]

ANSWER

Organizational Structure	*Project Manager Authority*
Functional	Little or no authority
Weak matrix	Limited authority
Balanced matrix	Low to moderate authority
Strong matrix	Moderate to high authority
Projectized	High to almost total authority

[Planning]

29

Which of the five methods of resolving conflict is recommended? Why?

30

In which two forms of project organization will an administrative staff generally be assigned to support a full-time project manager?

ANSWER

Problem solve or
confrontation.

Because both parties can be
fully satisfied if they work
together to find a solution that
satisfies both their needs.

[Executing]

ANSWER

- Strong matrix

- Projectized

[Planning]

31

QUESTION

What are the five types of power available to the project manager?

32

QUESTION

What is withdrawal or avoiding? What is its advantage and disadvantage in managing conflict?

ANSWER

- Legitimate
- Coercive
- Reward
- Expert
- Referent

[Executing]

ANSWER

Retreating from actual or potential disagreements and conflict situations.

Advantage: It cools the situation temporarily.

Disadvantage: It is a delaying tactic that fails to resolve the conflict.

[Executing]

33

Why is the type of organizational structure important in project management?

34

Give three examples of hygiene factors in Herzberg's theory of motivation. How do they affect motivation?

ANSWER

The structure of the performing organization often constrains the availability of, or terms under which, resources become available to the project.

The type of organizational structure also dictates the power and authority the project manager has to make decisions.

[Planning]

ANSWER

- Pay
- Attitude of supervisor
- Working conditions

Poor hygiene factors may destroy motivation, but improving hygiene factors are not likely to increase motivation.

[Planning]

35 QUESTION

How do the project coordinator and project expeditor organizational structures differ?

36 QUESTION

What type of organizational structure is represented in this figure?

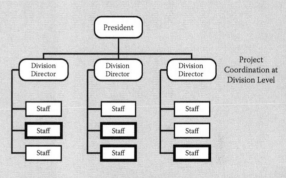

Project Coordination at Division Level

Bold boxes represent staff working on project activities.

ANSWER

The project coordinator reports to a higher-level manager than the project expeditor.

[Planning]

ANSWER

Functional organization.

[Planning]

37

Which type(s) of power should
a project manager use? Avoid?

38

Describe one of the primary
concerns of the project
manager in a projectized
organization.

ANSWER

Use reward and expert power; avoid using coercive power.

[Executing]

ANSWER

Ensuring that project team members are properly placed after the project is complete.

[Planning]

39

What are the four factors that influence the selection of the conflict resolution method?

40

Describe the difference between a weak matrix and a strong matrix.

ANSWER

- Relative importance and intensity of the conflict

- Time pressure to resolve the conflict

- Position taken by the people involved

- Motivation to resolve the conflict for a long-term or short-term basis

[Executing]

ANSWER

Weak matrices are similar to functional organizations (balance of power is tipped toward the functional manager). Strong matrices are similar to projectized organizations (balance of power is tipped toward the project manager).

[Planning]

41 QUESTION

Name the two outputs of the develop project team process.

42 QUESTION

What is a projectized organization?

ANSWER

- Team performance assessments
- Enterprise environmental factors updates

[Executing]

ANSWER

One in which a separate organization is established for each project. Personnel are assigned to particular projects on a full-time basis.

[Planning]

43

QUESTION

What is smoothing or accommodating? What are its advantage and disadvantage in resolving conflict?

44

QUESTION

What type of organizational structure is represented in this figure?

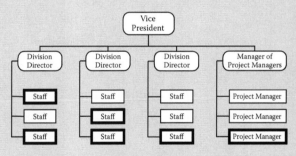

Bold boxes represent the project team.

ANSWER

Deemphasizing the opponents' differences and emphasizing their commonalities over the issues in question.

Advantage: It keeps the atmosphere friendly.

Disadvantage: It avoids solving the root causes of conflict.

[Executing]

ANSWER

Strong matrix organization.

[Planning]

45

What are the seven items in
the staffing management
plan?

46

When is the project team
directory prepared?

ANSWER

- Staff acquisition
- Resource calendars
- Staff release plan
- Training needs
- Recognition and rewards
- Compliance
- Safety

[Planning]

PROJECT **Human Resource** MANAGEMENT

ANSWER

As an output of the acquire project team process.

[Executing]

PROJECT **Human Resource** MANAGEMENT

47

QUESTION

What is compromising or reconciling? What are its advantage and disadvantage in resolving conflicts?

48

QUESTION

List six advantages for using a virtual team.

ANSWER

Bargaining and searching for solutions that attempt to bring some degree of satisfaction to the conflicting parties.

Advantage: Each party gets some degree of satisfaction.

Disadvantage: Neither party wins.

[Executing]

ANSWER

- Form teams of people who live in different geographic areas
- Include people with specialized skills who may live anywhere in the world
- Add people who work from home
- Ability to add people who work different shifts
- Include people who may have disabilities that prohibit them from traveling
- Launch projects in the face of tight travel budgets

[Executing]

49 QUESTION

What are the three examples of negotiation in the acquire project team process?

50 QUESTION

List seven interpersonal skills that a project manager should have or develop that are particularly important to team development.

ANSWER

With functional managers so the project receives competent staff as required, and team members are able, willing, and authorized to work on the project until their responsibilities are completed

With other teams in the organization to acquire scarce or specialized resources

With external organizations for appropriate scarce, specialized, qualified, certified, or other specified human resources

[Executing]

ANSWER

- Communications skills
- Emotional intelligence
- Conflict resolution
- Negotiation
- Influence
- Team building
- Facilitation

[Executing]

51

What is the definition of a
virtual team?

52

What are definitions for
authority and competency?
Where are they described?

A group of people with shared goals and objectives who fulfill their roles with little or no time spent meeting face to face.

[Executing]

Authority is the right to apply project resources, make decisions, sign approvals, accept deliverables, and carry out the work of the project.

Competency is the skill and ability required to complete assigned activities within project constraints.

They are included in the project's human resource management plan.

[Planning]

53

What are five stages of project
team development?

PROJECT **Human Resource** MANAGEMENT

54

How can project team
performance be improved?

PROJECT **Human Resource** MANAGEMENT

ANSWER

- Forming
- Storming
- Norming
- Performing
- Adjourning

[Executing]

ANSWER

By conducting formal or informal assessments of performance to pinpoint and resolve issues, manage conflict, and generally enhance team interaction.

[Executing]

55

QUESTION

In what circumstances are team members usually preassigned to a project?

56

QUESTION

What are eight examples of selection criteria that can be used to acquire team members?

ANSWER

If the project is the result of a competitive proposal and specific team members were identified in it, if the project depends on the expertise of particular people, or if some assignments are defined in the project charter.

[Executing]

ANSWER

- Availability
- Cost
- Experience
- Ability
- Knowledge
- Skills
- Attitude
- International factors

[Executing]

57

When is training most appropriate? What are six examples of training methods?

58

What are six examples of organizational process assets as an input to manage project team?

Aɴsᴡᴇʀ

Training is used to enhance the competencies of project team members

Examples are:

- Classroom
- Online
- Computer based
- On the job from another project team member
- Mentoring
- Coaching

[Executing]

Aɴsᴡᴇʀ

- Certificates of appreciation
- Newsletters
- Web sites
- Bonus structures
- Corporate apparel
- Other organizational prerequisites

[Executing]

59

QUESTION

What are seven guidelines for effective decision making?

60

QUESTION

Who has the primary responsibility for managing the dual reporting relationship between the team member and his or her functional manager and project manager?

A

ANSWER

- Focus on the goals to be served
- Follow a decision-making process
- Study environmental factors
- Analyze available information
- Develop personal qualities of the team members
- Stimulate team creativity
- Manage risk

[Executing]

A

ANSWER

The project manager.

[Executing]

1

Define Project
Communications
Management.

2

Name three tools and
techniques for control
communications.

ANSWER

The processes required to ensure timely and appropriate generation, collection, distribution, storage, retrieval, and ultimate disposition of project information.

[Planning, Executing, and Monitoring and Controlling]

PROJECT **Communications** MANAGEMENT

ANSWER

- Information management systems
- Expert judgment
- Meetings

[Monitoring and Controlling]

PROJECT **Communications** MANAGEMENT

3

What are three processes
in Project Communications
Management?

4

Why is planning
communications important to
project success?

ANSWER

- Plan Communications Management
- Manage Communications
- Control Communications

[Planning, Executing, and Managing and Controlling]

ANSWER

Inadequate communications planning may lead to problems that may include delay in message delivery, communicating to the wrong audience, insufficient communications to stakeholders, or misunderstanding or misinterpretation of the message communicated.

[Planning]

5 QUESTION

When should plan communications management be performed and why?

6 QUESTION

List 12 dimensions of communications.

ANSWER

As early in the project as possible as the project management plan is prepared.

The purpose is to have appropriate resources (time and budget) to be allocated to communications management activities.

[Planning]

ANSWER

- Internal/External
- Formal/Informal
- Vertical/Horizontal
- Official/Unofficial
- Written/Oral
- Verbal/Nonverbal

[Planning]

7 QUESTION

What is the difference
between effective and
efficient communications?

8 QUESTION

What are six important
considerations to take
into account in plan
communications?

Answer

Effective communications mean the information is provided in the right format, at the right time, to the right audience, and with the right impact.

Efficient communications mean providing only the information required.

[Planning]

Answer

- Who needs what information and who is authorized to access it?
- When do people need the information?
- Where will the information be stored?
- What format will be used to store the information?
- How will the information be retrieved?
- What time zones, language barriers, and cross-cultural considerations should be taken into account?

[Planning]

9

QUESTION

Give five examples of factors that will affect the type of communications technology used in any project.

10

QUESTION

List five tools and techniques in manage communications

PROJECT Communications MANAGEMENT

ANSWER

- Urgency of the need for the information
- Technology availability
- Ease of use
- Project environment
- Sensitivity and confidentiality of the information

[Planning]

ANSWER

- Communications technology
- Communications models
- Communications methods
- Information management systems
- Performance reporting

[Executing]

11

QUESTION

Describe the purpose of the communications management plan and the guidelines and templates it can include.

12

QUESTION

What are the four inputs to the manage communications process?

ANSWER

- It describes how project communications management will be planned, structured, monitored, and controlled.

- Guidelines and templates in it may include ones for project status meetings, project team meetings, e-meetings, and e-mail messages.

[Planning]

ANSWER

- Communications management plan

- Work performance information

- Enterprise environmental factors

- Organizational process assets

[Executing]

13

QUESTION

List seven pieces of
information typically
needed to determine
project communications
requirements.

14

QUESTION

How is communications
requirements analysis used
in plan communications
management?

ANSWER

- Organization charts

- Project organization and stakeholder responsibility relationships

- Disciplines, departments, and specialties involved in the project

- Number and locations of persons involved in the project

- External information needs

- Internal information needs

- Stakeholder information

[Planning]

ANSWER

As a tool and technique to determine the information needs of stakeholders; the requirements are defined by combining the type and format of information needed with an analysis of the value of the information. The purpose is to use project resources only on communicating information that contributes to project success.

[Planning]

15

What are the major responsibilities of the sender and receiver in communicating?

16

Provide two examples of project documents to update as an output of plan communications management:

ANSWER

Sender—

- Makes the information clear, unambiguous, and complete so that the receiver can receive it correctly

- Confirms that the information is properly understood

Receiver—

- Ensures that the information is received in its entirety

- Ensures that the information is understood correctly and acknowledged

[Planning]

ANSWER

- Project schedule
- Stakeholder register

17

What are four important
characteristics of an effective
issue log?

18

What are three examples of
project documents to update
as an output of manage
communications?

ANSWER

- Issues are clearly stated

- Issues are categorized based on urgency and potential impact

- An owner is assigned to every issues

- A target date is established for closure

[Monitoring and Controlling]

ANSWER

- Issue log

- Project schedule

- Funding requirements

[Executing]

19

QUESTION

How is the issue log used in control communications?

20

QUESTION

What are the five steps in the basic communications model? Describe who is responsible for each step.

A NSWER

As an input to facilitate communications and ensure a common understanding of issues. It helps and documents who is responsible for resolving issues by a specific date and addresses obstacles that may be a barrier to the team in terms of achieving its goals.

[Monitoring and Controlling]

A NSWER

- Encode: Sender translates thoughts or ideas into language
- Transmit message: Sender sends information using various communication channels
- Decode: Receiver translates the message back into meaningful thoughts or ideas
- Acknowledge: Receiver states receipt of the message but does not indicate agreement and/or comprehension of it
- Feedback/Response: Receiver includes thoughts and ideas into a message and transmits it back to the original sender

[Planning]

21

Why are updates to the project management plan an output of manage communications?

22

What is performance reporting? Where is it used in communications management

Aɴsᴡᴇʀ

The project management plan provides information on project baselines.

Communications management, and stakeholder management, which may require updates on current project performance against the performance measurement baseline.

[Executing]

Aɴsᴡᴇʀ

Collecting and distributing performance information such as status reports, process measurements, and forecasts.

It is used as a tool and technique in manage communications.

[Executing]

23

QUESTION

While a simple status report shows performance information or sample dashboards for scope, schedule, cost, or quality, list seven items that are included in more detailed reports.

24

QUESTION

What are three communications methods to share information with project stakeholders?

ANSWER

- Analysis of past performance
- Analysis of project forecasts including time and cost
- Current status of risks and issues
- Work completed during the period
- Work to be completed in the next period
- Summary of changes approved in the period
- Any other relevant information that is reviewed and discussed

[Executing]

ANSWER

- Interactive communication
- Push communication
- Pull communication

[Planning]

25

What is the most efficient way to ensure a common understanding by all participants on specific topics, and how is it done?

26

Describe the two most important organizational process assets used in planned communications management and why you selected them.

ANSWER

Interactive communication between two or more parties in a multidirectional exchange of information. It is handled through meetings, phone calls, instant messages, and video conferences.

[Planning]

ANSWER

Lessons learned and historical information.

They can provide insight into the decisions made regarding communications issues and results of these decisions on previous projects as guiding information to plan the project's communication activities.

[Planning]

27

What is the purpose of the manage communications process?

28

What is active listening?

ANSWER

The process of monitoring and controlling communications throughout the entire project life cycle to ensure the project stakeholders' information needs are met.

[Monitoring and Controlling]

ANSWER

Listening in which the recipient is attentive in terms of acknowledging, clarifying, and confirming understanding and removing any barriers that could affect adversely comprehension.

[Executing]

29

What are two examples
of meeting management
techniques?

30

What are two examples of
facilitation techniques?

ANSWER

Preparing an agenda and
dealing with conflicts.

[Executing]

ANSWER

Building consensus and
overcoming obstacles.

[Executing]

31

Why is it important that project managers develop a sensitivity to nonverbal messages?

32

What are two key items to consider in presentation techniques?

ANSWER

Because studies have shown that nonverbal cues are a better indicator of the meaning behind the message than the words used.

[Planning]

ANSWER

Awareness of the impact of body language and design of visual aids.

[Executing]

33

QUESTION

What is a project "war room"?
What is its primary benefit?

34

QUESTION

What are three items to
consider in terms of writing
style?

ANSWER

A single location for the project team to get together for any purpose. The war room should provide a repository for project artifacts, records, and up-to-date schedules and status reports.

Gives identity to the project team.

[Planning]

ANSWER

Active versus passive voice, sentence structure, and word choice.

[Executing]

35

What are three items to consider in terms of choice of communications media?

36

Why is plan communications management tightly linked to enterprise environmental factors?

ANSWER

When to use written communications versus oral communications, when to use an informal memo versus a formal project report, and when face-to-face communication should be used versus e-mail.

[Executing]

ANSWER

Because the project's organizational structure will have a major effect on project communications.

[Planning]

37

What are two ways to enhance sender-receiver models?

38

What are six examples of organizational process assets used in control communications?

ANSWER

By incorporating feedback loops to provide opportunities for interactive participation and by removing communications barriers.

[Executing]

ANSWER

- Report templates
- Policies, standards, and procedures to define communications
- Specific communication technologies available
- Allowed communication media
- Record retention policies
- Security requirements

[Monitoring and Controlling]

39

How can work performance
reports best be used to
manage communications?
How can their use be
optimized?

40

What are three enterprise
environmental factors that
can influence manage
communications?

ANSWER

To collect project performance and status information to facilitate discussion and create communications. These reports should be comprehensive, accurate, and available in a timely way to optimize their use.

[Executing]

ANSWER

- Organizational culture and standards

- Government or industry standards or regulations

- Project management information system

[Executing]

41

What are three examples of organizational process assets that can influence manage communications?

42

What are five inputs to the control communications process?

ANSWER

- Communication management policies, procedures, processes, and guidelines

- Templates

- Historical information and lessons learned

[Executing]

ANSWER

- Project management plan

- Project communications

- Issue log

- Work performance information

- Organizational process assets

[Monitoring and Controlling]

43

How is push communications
used? Provide eight examples.

44

What are four outputs of
manage communications?

A NSWER

To send specific recipients information and ensure it is distributed.

Examples:
- Letters
- Memos
- Reports
- Emails
- Faxes
- Voice mails
- Blogs
- Press releases

[Planning]

A NSWER

- Project communications
- Project management plan updates
- Project document updates
- Organizational process assets updates

[Executing]

45

QUESTION

How is push communications used? Provide four examples.

46

QUESTION

How is an information management system used in control communications?

ANSWER

For very large volumes of information or for information to very large audiences; recipients access information at their own discretion.

Examples:

- Intranet sites

- E-Learning

- Lessons learned databases

- Knowledge repositories

[Planning]

ANSWER

As a tool and technique to provide standard tools for the project manager to capture, store, and distribute information to stakeholders about the project's costs, schedule progress, and performance.

[Monitoring and Controlling]

47

QUESTION

What formula is used to calculate the number of communication channels on a project?

PROJECT **Communications** MANAGEMENT

48

QUESTION

Describe five types of valuable information in the project management plan that is useful to the control communications process.

PROJECT **Communications** MANAGEMENT

ANSWER

n * (n-1) / 2

n = [the number of stakeholders]

[Planning]

ANSWER

- Stakeholder communication requirements

- Reason for the information distributed

- Timeframe and frequency for the distribution of the required communication

- Individual or group responsible for communicating the information

- Individual or group receiving the information

[Monitoring and Controlling]

49

How would you determine
the choice of communications
methods to use on your
project?

50

What are three examples of
information management
systems?

ANSWER

By discussions and reaching agreement with stakeholders based on communication requirements, cost and time constraints, and familiarity and availability of required tools and resources for communications.

[Planning]

ANSWER

- Hard copy document management

- Electronic communications management

- Electronic project management tools

[Executing]

51

QUESTION

How is the performance
management baseline
affected by the manage
communications process?

52

QUESTION

What is the key benefit of
control communications?

ANSWER

An output of the process is updates to the project management plan. It provides information on baselines, communication management, and stakeholder management. Each of these areas may require updates based on the current project performance versus the project management baseline.

[Executing]

ANSWER

To ensure an optimal information flow among all communication participants at any moment of time.

[Monitoring and Controlling]

53

Define the performance management baseline.

54

What are two examples of communication elements that would immediately trigger a revision to the plan communications management or manage communications processes?

Aɴsᴡᴇʀ

It is an approved plan for the project work to which project execution is compared, and deviations are measured for project control.

[Executing]

Aɴsᴡᴇʀ

Issues

Key performance indicators such as actual versus planned schedule, cost, or quality.

[Planning, Executing, and Monitoring and Controlling]

55

QUESTION

What five items or parameters may be integrated in the performance management baseline?

56

QUESTION

How is expert judgment used in control communications?

ANSWER

- Scope
- Schedule
- Cost
- Technical
- Quality

[Executing]

ANSWER

As a tool and technique to assess the impact of the project communications, need for action or intervention, action that should be taken, responsibility for taking the action, and the required time frame for the action.

[Monitoring and Controlling]

57

List seven types of people or groups that can provide expert judgment in control communications:

58

Who is responsible for determining the actions required to ensure that the right message is communicated to the right audience at the right time?

ANSWER

- Other units in the organization
- Consultants
- Stakeholders, including customers and sponsors
- Professional and technical associations
- Industry groups
- Subject matter experts
- The project management office

[Monitoring and Controlling]

ANSWER

The project manager in collaboration with the project team

[Monitoring and Controlling]

59

How are change requests used in manage communications?

60

What are four possible actions that may result from change requests from manage communications?

ANSWER

As an output of the process because there may be a need for adjustment, intervention, or action.

[Executing]

ANSWER

- Provide new or revised cost estimates, activity sequences, schedule dates, resource requirements, and analysis of risk response alternatives

- Adjust the project management plan and other documents

- Recommend corrective actions to bring future performance in line with the project management plan

- Recommend preventive action to reduce the probability of incurring future negative project performance

[Monitoring and Controlling]

1

What is Project Risk Management?

PROJECT **Risk** MANAGEMENT

2

How is Monte Carlo analysis used in schedule simulations?

Aɴꜱᴡᴇʀ

The processes concerned with planning, identifying, analyzing, responding, and monitoring and controlling project risk. It also includes maximizing the results of positive events as well as minimizing the consequences of adverse events to project objectives.

[Planning, Monitoring and Controlling]

Aɴꜱᴡᴇʀ

To "perform" the project many times in order to provide a statistical distribution of the calculated results. For a schedule risk analysis, the schedule network diagram and duration estimates are used.

[Planning]

3

QUESTION

Generally speaking, in what project phase are risk and opportunity greater, by the widest margin, than the amount at stake?

4

QUESTION

What is a reserve?

ANSWER

Concept phase.

[Planning]

ANSWER

A provision (usually time or money) in the project management plan to mitigate cost and/or schedule risk. Typically, there are two reserves on any project: management reserve and contingency reserve.

[Planning]

5

QUESTION

What is the purpose of a
risk rating in the perform
qualitative risk analysis
process?

6

QUESTION

Calculate the probability and
expected monetary value
(EMV) of path A–C of the
figure below.

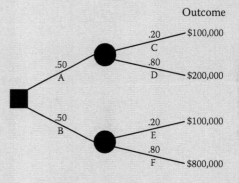

Outcome

	.20	$100,000
	C	
.50		
A	.80	
	D	$200,000

	.20	$100,000
	E	
.50		
B	.80	
	F	$800,000

ANSWER

To focus the project team's efforts on developing responses for those risks that have the highest probability of impact and occurrence.

[Planning]

ANSWER

Probability = (.50) (.20) = 10%

EMV = (.50) (.20) ($100,000) = $10,000

[Planning]

7

QUESTION

How is the risk impact scale used?

8

QUESTION

A project has a 50 percent chance of making a $10,000 profit and a 20 percent chance of losing $50,000. What is the EMV of the project?

ANSWER

To identify the severity of each risk's effect on the project objectives.

[Planning]

ANSWER

EMV = (.50) ($10,000) + (.20) (–$50,000) = –$5,000

[Planning]

9

QUESTION

What are triggers? What other term can be used to define them? Provide an example.

10

QUESTION

What is risk mitigation?

ANSWER

An indication or warning sign that a risk has occurred or is about to occur.

Risk symptom.

Failure to meet intermediate milestones may be an early warning signal of an impending schedule delay.

[Planning]

ANSWER

A risk response strategy is where the project team acts to reduce the probability of occurrence or impact if the risk. Taking early action to reduce the probability and/or impact of a risk occurring on the project is often more effective than trying to repair the damage after the risk has occurred.

It is a tool and technique in plan risk responses.

[Planning]

11

What is the purpose of a risk management plan? List five of the ten areas it should include.

12

How can earned value be used in control risks?

ANSWER

To describe to all stakeholders how risk management will be addressed and performed on the project.

Methodology; roles and responsibilities; budgeting; timing; risk categories; definitions of risk probability and impact; probability and impact matrix; revised stakeholder tolerances; reporting formats; and tracking.

[Planning]

ANSWER

To monitor overall project performance by indicating potential deviation of the project at completion from cost and schedule targets. Deviation may indicate the potential impact of threats or opportunities.

[Monitoring and Controlling]

13

QUESTION

What is risk appetite?

14

QUESTION

What is the purpose of reserve analysis?

ANSWER

The degree of uncertainty an entity is willing to take on in anticipation of a reward.

[Planning]

PROJECT **Risk** MANAGEMENT

ANSWER

To compare contingency reserves remaining against risk remaining in order to determine if the remaining reserve is adequate.

[Monitoring and Controlling]

PROJECT **Risk** MANAGEMENT

15

Describe the output from identify risks.

16

What is statistical independence?

ANSWER

Risk register, which will now include the following initial entries:

- List of identified risks

- List of potential responses

[Planning]

ANSWER

The independence of two events in which the occurrence of one is not related to the occurrence of the other.

[Planning]

17

When is a contingent response strategy used?

18

What are four ways to deflect or transfer a risk?

ANSWER

Only when certain, previously identified, events occur. Events that trigger the contingency response should be defined and tracked.

[Planning]

PROJECT **Risk** MANAGEMENT

ANSWER

- Purchase insurance or a performance bond
- Use a warranty
- Use a guarantee
- Use a subcontractor

[Planning]

PROJECT **Risk** MANAGEMENT

19

What is risk acceptance?

20

How are checklists used in risk identification? Provide one advantage and one disadvantage of using a checklist.

ANSWER

Acknowledging a risk but taking no action to avoid, transfer, or mitigate it. Acceptance can be active (for example, preparing a contingency plan) or passive (that is, dealing with the risk if and when it occurs). The risk may be a threat or an opportunity.

[Planning]

ANSWER

To itemize all types of possible risk to the project.

Advantage: Quick and simple.

Disadvantage: Impossible to build an exhaustive checklist.

[Planning]

21

What are four strategies for dealing with opportunities?

22

During which process are risks prioritized for further analysis or action by assessing and combining their probability of occurrence and impact?

ANSWER

- Exploit
- Share
- Enhance
- Accept

[Planning]

ANSWER

Perform qualitative risk analysis.

[Planning]

23

What are risk consequences?

24

At which phase(s) in the project life cycle is significant risk thought to have the greatest impact?

The effect on project
objectives if the risk event
occurs.

[Planning]

PROJECT **Risk** MANAGEMENT

Project Execution and
Closeout (Termination).

[Planning]

PROJECT **Risk** MANAGEMENT

25

In the following figure, what is the probability that Milestone B will be completed according to schedule, given that activities 1, 2, and 3 each have a 90 percent probability of completion according to schedule?

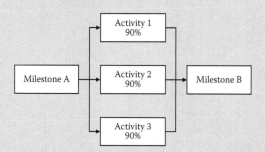

26

What types of enterprise environmental factors are inputs to identify risks?

ANSWER

Probability = (.90) (.90) (.90)
$$= 73\%$$

[Planning]

ANSWER

- Commercial databases
- Academic studies
- Published checklists
- Benchmarking
- Industry studies
- Risk attitudes

[Planning]

27

QUESTION

What is a project risk?

28

QUESTION

What is the difference
between a risk and a problem?

PROJECT **RISK** MANAGEMENT

PROJECT **Risk** MANAGEMENT

ANSWER

An uncertain event that, if it occurs, has a positive or negative effect on at least one project objective, such as time, cost, scope, or quality.

[Planning]

ANSWER

A risk is a future phenomenon (or event); it has not yet occurred. A problem, on the other hand, currently exists. Problems are often called issues.

[Planning]

PROJECT **Risk** MANAGEMENT

PROJECT **Risk** MANAGEMENT

29

What should you do to avoid
a risk?

30

When should a risk urgency
assessment be performed?
What are four indicators of
priority?

Adopt an alternative strategy—find another approach to get the job done, which entails changing the project management plan.

[Planning]

During perform qualitative risk analysis

- Probability of detecting the risk
- Time to effect a risk response
- Symptoms and warning signs
- Risk rating

[Planning]

31 QUESTION

How are contingency reserves calculated?

32 QUESTION

What is a risk premium?

ANSWER

Based on the quantitative analysis of the project and the organization's risk thresholds.

[Planning]

ANSWER

Money paid to transfer the risk to a third party.

[Planning]

33 QUESTION

What is the responsibility of the risk response owner in the control risks process?

34 QUESTION

This figure presents results obtained using what technique?

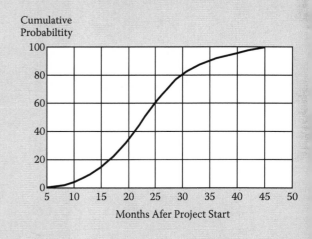

ANSWER

To periodically report to the project manager—

- The effectiveness of the risk plan

- Unanticipated effects of the risk under observation

- Any corrective action taken

[Monitoring and Controlling]

ANSWER

Monte Carlo simulation of a project schedule.

[Planning]

35

QUESTION

What is a risk breakdown structure?

36

QUESTION

What are four updates from the perform qualitative risk analysis process?

ANSWER

A hierarchical depiction of all identified project risks structured by risk category and subcategory that pinpoints the areas and causes of potential risk. It is one of several techniques used in risk management plan.

[Planning]

ANSWER

- Assessments of probability and impacts for each risk
- Risk ranking or scores
- Risk urgency information or risk categorization
- Watch list for low probability risks or risks requiring further action

[Planning]

37

QUESTION

What is used as a model
for simulation for a cost risk
analysis?

38

QUESTION

Risk attitudes of the
organization and stakeholders
can be influenced by three
factors, name them.

ANSWER

Cost estimates

[Planning]

ANSWER

- Risk appetite
- Risk tolerance
- Risk threshold

[Planning]

39

QUESTION

Who should attend a meeting to develop the risk management plan?

40

QUESTION

This figure represents what risk quantification technique?

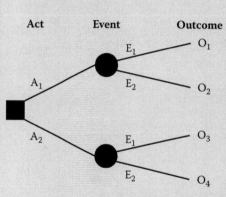

ANSWER

- Project manager
- Selected project team members
- Anyone in the organization responsible for managing the risk planning and execution activities
- Stakeholders

[Planning]

ANSWER

Decision-tree analysis.

[Planning]

41

QUESTION

When should you avoid a
risk? Provide four examples of
avoidance.

42

QUESTION

What is the output of the
perform quantitative risk
analysis process?

- When you want to eliminate it.

Examples include:

- Extending the schedule
- Changing the strategy
- Reducing scope
- Shutting down the project entirely

[Planning]

PROJECT **Risk** MANAGEMENT

Project documents updates

[Planning]

PROJECT **Risk** MANAGEMENT

43

How is expected monetary value computed?

44

What does a continuous probability distribution represent? What are two examples?

ANSWER

Expected monetary value is the product of two numbers: risk event probability (an estimate of the probability that a given risk event will occur) and risk event value (an estimate of the gain or loss that will be incurred if the risk does occur).

[Planning]

ANSWER

The uncertainty in values, such as durations of schedule activities and costs of project components.

Beta and triangular distributions.

[Planning]

45

What are three categories of risk?

46

When should identify risks be performed?

ANSWER

- Sources of risk
- The area of the project affected
- Common root causes

[Planning]

ANSWER

Identify risks is not a one-time event; it should be performed on a regular basis throughout the project.

[Planning]

47

What is a decision tree? How is it used in perform quantitative risk analysis?

48

What is a sensitivity analysis? What is a typical display of a sensitivity analysis?

ANSWER

A diagram that shows a decision that is being considered and the implications of choosing one or more available alternatives. The branches of the tree represent either decisions (shown as boxes) or chance events (shown as circles).

Used to compute expected monetary values for each branch of the tree to help the project team quantify each decision alternative.

[Planning]

ANSWER

A method used to determine which risks have the greatest potential impact on the project.

Tornado diagram.

[Planning]

49

Electing to use older technology rather than buying the latest device represents what type of risk response?

50

How are schedule simulation results used?

ANSWER

Avoidance.

[Planning]

ANSWER

To quantify the risk of various schedule alternatives, different project strategies, different paths through the network, or individual activities.

[Planning]

51

QUESTION

List the six processes in project risk management?

52

QUESTION

What are residual risks?

- Plan risk management
- Identify risks
- Perform qualitative risk analysis
- Perform quantitative risk analysis
- Plan risk responses
- Control risks

[Planning and Monitoring and Controlling]

PROJECT **Risk** MANAGEMENT

A risk that remains after responses have been implemented.

[Planning]

PROJECT **Risk** MANAGEMENT

53

What are secondary risks?

54

What is the most common active risk acceptance strategy?

ANSWER

Risks that arise after
implementing a risk response

[Planning]

ANSWER

Establishing a contingency
reserve, to include amounts of
time, money, or resources, to
handle the risks, should they
occur.

[Planning]

55 QUESTION

What is the purpose of risk audits? When are they performed?

56 QUESTION

For what category of risk would you use the "exploit" strategy?

ANSWER

To assess the effectiveness of the project's risk management process and to examine and document the effectiveness of planned or unplanned risk responses in dealing with risk occurrences.

Throughout the project life cycle.

[Monitoring and Controlling]

ANSWER

For positive risks or opportunities.

[Planning]

57

QUESTION

What is a SWOT analysis?
How is it used?

58

QUESTION

What is an assumptions
analysis? How is it used?

ANSWER

A strengths, weaknesses, opportunities, and threats analysis.

Used in identify risks, it tends to increase the number of project risks considered.

[Planning]

ANSWER

An examination of the validity of assumptions as they apply to the project.

Used in identify risks to identify risks by examining the inaccuracy, inconsistency, or instability or incompleteness of assumptions. Assumptions are tested during perform qualitative risk analysis.

[Planning]

59

QUESTION

What action should the
project manager take
to ensure that each risk
response is implemented and
monitored?

60

QUESTION

What are three examples of
risk diagramming techniques?

ANSWER

Identify and assign an individual to own each risk who will take personal responsibility for protecting the project objectives from the risk's impact.

[Planning]

ANSWER

- Cause-and-effect diagrams
- Systems or process flowcharts
- Influence diagrams

[Planning]

1

QUESTION

Name the four processes
involved in Project
Procurement Management.

2

QUESTION

List the four tools and
techniques for the plan
procurement management
process.

ANSWER

- Plan procurement management
- Conduct procurements
- Control procurements
- Close procurements

[Planning, Executing, Monitoring and Controlling, and Closing]

ANSWER

- Make-or-buy analysis
- Expert judgment
- Market research
- Meetings

[Planning]

3

QUESTION

What type of change
constitutes one of the major
areas of cost growth?

4

QUESTION

What is the purpose of a
warranty?

ANSWER

Change to project scope.

[Monitoring and Controlling]

ANSWER

As stated in source selection criteria, it states what the seller proposes to warrant for the final product and through what time period.

[Planning]

5

List the nine inputs to the plan procurement management process.

6

List the four other project management processes that may be applied to the control procurement process.

PROJECT **Procurement** MANAGEMENT

ANSWER

- Procurement management plan
- Requirements documentation
- Risk register
- Activity resource requirements
- Project schedule
- Activity cost estimates
- Stakeholder register
- Organizational process assets
- Enterprise environmental factors

[Planning]

ANSWER

- Direct and manage project work
- Control quality
- Perform integrated change control
- Control risks

[Monitoring and Controlling]

7
QUESTION

What is a make-or-buy
analysis?

8
QUESTION

Why should advertising
be used in the conduct
procurements process?

ANSWER

A technique used to
determine whether particular
work can be accomplished by
the project team or must be
purchased from an outside
source.

[Planning]

ANSWER

To expand existing lists of
potential sellers.

[Executing]

9

QUESTION

What is accomplished during conduct procurements?

10

QUESTION

Who bears the cost risk in a cost-reimbursable contract—the seller or the buyer? Explain.

ANSWER

Seller responses are obtained, sellers are selected, and contracts are awarded.

[Executing]

ANSWER

The buyer.

He or she is obligated to reimburse the seller for all allowable and allocable costs incurred during contract performance.

[Planning]

11

Provide two examples of
requirements documentation
as an input to plan
procurement management.

PROJECT **Procurement** MANAGEMENT

12

What are the objectives of
procurement performance
reviews?

ANSWER

Information about project requirements.

Requirements with contractual and legal implications that may include health, safety, security, performance, environmental insurance, intellectual property rights, equal employment opportunity, licenses, and permits.

[Planning]

ANSWER

To identify performance successes or failures, progress with respect to the procurement statement of work, and contract noncompliance, which would allow the buyer to quantify the seller's demonstrated ability or inability to perform work.

[Monitoring and Controlling]

13

What are teaming agreements? How do they impact the roles of the buyer and seller?

14

Describe a time-and-materials contract.

ANSWER

Legal contractual agreements between two or more parties to form a partnership or joint venture, or similar relationship.

The agreement defines buyer-seller roles for each party so the roles are predetermined.

[Planning]

ANSWER

A contractual arrangement that contains aspects of both cost-reimbursable and fixed-price arrangements. Time-and-materials contracts are open ended to the extent that the full value of the arrangement is not defined at the time of the award; but they are fixed to the extent that unit rates are preset by the buyer and seller.

[Planning]

15

QUESTION

Where are the requirements
for formal procurement
closure found?

16

QUESTION

What is a cost-plus-fixed-fee
contract? When do the fee
amounts change?

ANSWER

In the contract and the procurement management plan.

[Planning]

ANSWER

A contract under where the seller is reimbursed for all allowable costs and a fixed fee calculated as a percentage of the initial estimated contract cost. The fee is paid only for completed work and only changes if the project scope changes.

[Planning]

17

QUESTION

What are the characteristics of a well-written SOW?

18

QUESTION

What is a cost-plus-incentive-fee contract? How are costs shared between buyer and seller?

ANSWER

The SOW should be clear, complete, and concise and should include a description of any collateral services required, such as performance reporting or post-contract operational or maintenance support for the procured item.

[Planning]

ANSWER

A contract where the seller is reimbursed for allowable costs and paid a predetermined incentive fee as a bonus for achieving stated performance objectives. If the final costs are less than or greater than the original estimated costs, both the buyer and seller share the differences based on a prenegotiated cost-sharing formula.

[Planning]

19

What is a fixed price with economic price adjustment contract?

20

What is the seller's risk objective in Project Procurement Management? What two broad categories of contract types satisfy this objective?

ANSWER

One that is used when the seller's performance period spans a number of years. It is a fixed-price contract that includes a special provision allowing for predetermined final adjustments to the contract price that are due to changed conditions such as inflation or cost increases or decreases for specific commodities.

[Planning]

ANSWER

To minimize risk while maximizing profit potential.

- Fixed-price contracts
- Time and material contracts

[Planning]

21

QUESTION

What is the key benefit of
the conduct procurements
process?

22

QUESTION

Why is the scope statement
an input to plan procurement
management?

ANSWER

It provides alignment of internal and external stakeholder expectations through established agreements.

[Executing]

ANSWER

Because it includes the product, service, or results description; the list of deliverables; acceptance criteria; and important information on technical issues or concerns that could impact cost estimating.

[Executing]

23

What is the difference between these four terms: bid, quotation, tender, and proposal?

24

List the three broad categories of contracts.

ANSWER

Bid, quotation, or tender generally are used if the seller selection decision will be based on price, while a proposal is used when other conditions such as technical capability and technical approach are significant.

[Planning]

ANSWER

- Cost reimbursement
- Time and materials
- Fixed price

[Planning]

25

QUESTION

Provide at least seven examples of source selection criteria used in plan procurement management.

26

QUESTION

What are seller performance evaluations? When are they prepared?

ANSWER

- Understanding of need
- Overall or life-cycle cost
- Technical capability
- Risk
- Management approach
- Technical approach
- Warranty
- Financial capacity
- Production capacity and interest
- Business size and type
- Past performance of sellers
- References
- Intellectual property rights
- Proprietary rights

[Planning]

ANSWER

Assessments of the seller's ability to continue to perform on the current contract; opinions on whether the seller should be allowed to perform work on future contracts or to rate how well the seller is performing on the contract.

As an output to control procurements

[Monitoring and Controlling]

27

What are qualified seller lists?
What is their purpose?

28

What is meant by requested
but unresolved changes?
How should they be handled?

ANSWER

Prequalified lists of sellers who have been prescreened for their qualifications and past experience. (To direct procurements to those sellers who can perform on resulting contracts.)

[Executing]

ANSWER

They can include direction provided by the buyer or actions by the seller that the other party considers a constructive change to the contract.

Since they may be disputed and may lead to a claim, they should be identified and documented in the project correspondence.

They require change requests.

[Monitoring and Controlling]

29

QUESTION

What is a cost-plus-award fee contract?

30

QUESTION

What is a bidder conference? When should it be held?

Answer

A contract in which the seller is reimbursed for all allowable costs, but the majority of the fee is earned based only on satisfying certain broad, subjective performance criteria defined and incorporated into the contract. The buyer determines the fee based on subjective determination of seller performance.

[Planning]

Answer

A bidder conference (also called a contractor conference, vendor conference, or prebid conference) is a meeting with prospective sellers to ensure that they have a clear, common understanding of the buyer's needs and contract terms and conditions.

Before sellers prepare their proposals.

[Executing]

31

QUESTION

What is the purpose of the procurement statement of work (SOW)? List seven items that may be included in it.

32

QUESTION

What is a fixed-price-incentive-fee contract?

ANSWER

It describes the procurement deliverables in sufficient detail so prospective sellers can determine if they can provide what the buyer has specified.

- Specifications
- Quantity desired
- Quality levels
- Performance data
- Period of performance
- Work locations
- Other requirements

[Executing]

ANSWER

One where the buyer pays the seller a fixed amount, as defined by the contract, and the seller can earn an additional amount if defined performance criteria are met.

[Planning]

33

What is a weighting system?
Why is it used in the conduct
procurements process?

34

What types of costs are
considered in a make-or-buy
analysis?

PROJECT **Procurement** MANAGEMENT

PROJECT **Procurement** MANAGEMENT

ANSWER

A method to quantify qualitative data as a proposal evaluation technique. It may be used to select a single seller to sign a standard contract or to establish a negotiating sequence by ranking proposals by the weighted evaluation scores.

[Executing]

ANSWER

Both the direct and indirect support costs of a prospective procurement.

[Planning]

35

What are make-or-buy decisions? When are they prepared?

36

List three administrative activities that are part of close procurements?

A

ANSWER

They are the result of a make-or-buy analysis to determine if the project team will do the work or if an outside source will be used. If the decision is to make, the procurement management plan may define internal processes and agreements. If the decision is to buy, a process is followed to reach an agreement with a seller.

They are an output of plan procurement management.

[Planning]

A

ANSWER

- Finalizing open claims
- Updating records to reflect final results
- Archiving this information for future use

[Closing]

37

What is the buyer's risk
objective in project
procurement management?

38

What is a contract?

ANSWER

To place maximum performance risk on the seller while maintaining incentives for economic and efficient performance.

[Planning]

ANSWER

A mutually binding agreement that obligates the seller to provide a specific service or product and obligates the buyer to pay for it; a legal relationship subject to remedy in court.

[Executing]

39

QUESTION

When is a short list of qualified sellers established? Why is this done?

40

QUESTION

What is control procurements?

ANSWER

In the conduct procurements process.

To accelerate source selection by negotiating only with sellers who have a reasonable chance of award.

[Executing]

ANSWER

The process of managing procurement relationships, monitoring contract performance, and making changes and corrections if needed.

[Monitoring and Controlling]

41

QUESTION

How is market research
used in plan procurement
management?

42

QUESTION

Why should unresolved claims
be carefully documented?

ANSWER

As a tool and technique to examine industry and vendor capabilities. The intent is to help refine particular procurement objectives to leverage maturing technologies, while balancing risks associated with the breadth of vendors who can provide the needed materials or services.

[Planning]

ANSWER

Because they may be subject to litigation.

[Closing]

43

QUESTION

When is the invitation for bid most appropriate?

44

QUESTION

How is performance reporting used as a tool and technique in control procurements?

ANSWER

For routine items when the primary objective is to award the contract based on the lowest price.

[Planning]

ANSWER

It provides management with information about how effectively the seller is achieving the contractual objectives.

[Monitoring and Controlling]

45

What is a firm-fixed-price contract? When is it used?

46

How does the close procurements process support the close project process?

ANSWER

The type of contract where the seller furnishes goods or services at a fixed price regardless of its costs.

It is best suited for situations where the buyer can describe what it needs using detailed specifications.

[Planning]

ANSWER

It verifies that all work and deliverables were acceptable.

[Closing]

47

In a fixed-price-incentive-fee (FPIF) contract, who is responsible for any costs that exceed the price ceiling?

48

List three reasons for early contract terminations.

ANSWER

The seller

[Planning]

ANSWER

- Mutual agreement of both parties
- Default of one party
- Convenience of the buyer if provided for in the contract

[Closing]

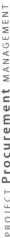

49 QUESTION

How are Internet searches used in conduct procurements?

50 QUESTION

What is a negotiated settlement?

ANSWER

As a tool and technique to communicate solicitations to the vendor community.

[Executing]

ANSWER

The final equitable settlement of all outstanding issues, claims, and disputes. Direct negotiation is the preferred method to reach a settlement; however, mediation or arbitration may also be considered. If all else fails, litigation is the last resort.

As a tool and technique in close procurements.

[Closing]

51

What is the purpose of the procurement management plan?

PROJECT **Procurement** MANAGEMENT

52

What is a procurement audit? How is it used?

PROJECT **Procurement** MANAGEMENT

ANSWER

To describe how the procurement processes will be managed, from developing procurement documentation through contract closure.

[Planning]

ANSWER

Structured review assessment and evaluation of the procurement process from plan procurements through control procurements.

Used to identify successes and failures which can be applied as lessons learned to other projects.

[Closing]

53

QUESTION

What type of procurement
documentation should
be reviewed in close
procurements?

54

QUESTION

What are two outputs from
close procurements?

ANSWER

Information on contract schedule, scope, quality, and cost performance; contract change documentation; payment records; and inspection results.

[Closing]

ANSWER

- Closed procurements
- Organizational process assets (updates)

[Closing]

55 QUESTION

What are examples of organizational process assets to update during control procurements?

56 QUESTION

What is a contract change control system? What does it include?

ANSWER

- Correspondence
- Payment schedules and requests
- Seller performance evaluation documentation

[Monitoring and Controlling]

ANSWER

Defines the process by which the procurement can be modified.

Includes paperwork, tracking systems, dispute resolution procedures, and approval levels necessary for authorizing changes.

[Monitoring and Controlling]

57

What are contested changes and potential constructive changes? What are three other terms for them?

58

How are independent estimates used in the conduct procurements process?

ANSWER

Changes in which the buyer and seller cannot reach an agreement on compensation for the change or cannot agree that a change has occurred.

They may be called claims, disputes, or appeals.

[Monitoring and Controlling]

PROJECT **Procurement** MANAGEMENT

ANSWER

To serve as a benchmark for proposed responses as significant differences in cost estimates may indicate the procurement statement of work was deficient, ambiguous, and/or prospective sellers have misunderstood or failed to respond to the procurement statement of work.

[Executing]

PROJECT **Procurement** MANAGEMENT

59

QUESTION

What are 10 items that may be part of procurement negotiations and then should be reflected in the final contract language?

60

QUESTION

List three organizational process assets to update as an output of close procurements.

ANSWER

- Responsibilities
- Authority to make changes
- Applicable terms and governing law
- Technical and business management approaches
- Technical solutions
- Overall schedule
- Payments
- Proprietary rights
- Contract pricing
- Price

[Executing]

ANSWER

- Procurement file
- Deliverable acceptance
- Lessons learned documentation

[Closing]

PROJECT **Stakeholder** MANAGEMENT

1

QUESTION

Define a stakeholder.

2

QUESTION

List the four processes in Project Stakeholder Management.

ANSWER

An individual, group, or organization that may be affected by or perceive itself to be affected by a decision, activity, or outcome of a project.

[Initiating and Planning]

PROJECT **Stakeholder** MANAGEMENT

ANSWER

- Identify stakeholders
- Plan stakeholder management
- Manage stakeholder engagement
- Control stakeholder engagement

[Initiating, Planning, Executing, and Monitoring and Controlling]

PROJECT **Stakeholder** MANAGEMENT

3

QUESTION

What is the benefit of plan
stakeholder management?

4

QUESTION

List three tools and techniques
in manage stakeholder
engagement.

ANSWER

It provides a clear, actionable path to interact with project stakeholders to support the interests of the project.

[Planning]

ANSWER

- Communication methods
- Interpersonal skills
- Management skills

[Executing]

5

QUESTION

How is an information management system used in control stakeholder and engagement? Provide two examples.

6

QUESTION

List four ways in which stakeholders can affect projects.

ANSWER

As a tool and technique

Examples:

- To capture, store, and distribute information to stakeholders about the project's cost, schedule progress, and performance.

- To enable the project manager to consolidate reports from several systems and facilitate distribution to stakeholders.

[Monitoring and Controlling]

ANSWER

- They may have active involvement in the project.

- They may have interests that may be positively or negatively affected by project completion.

- They may have competing expectations that may create conflicts on the project.

- They may exert influence over the project, its deliverables, and the team to achieve outcomes that satisfy strategic objectives or other needs.

[Initiating, Planning, Executing, and Monitoring and Controlling

7

QUESTION

Define plan identification
stakeholder.

8

QUESTION

List three tools and techniques
in plan stakeholder
identification.

ANSWER

The process of identifying the people, groups, or organizations that could be impacted by a decision, activity, or outcome of the project; and analyzing and documenting information on their interests, involvement, interdependencies, influence, and potential impact on project success.

[Initiating]

ANSWER

- Expert judgment
- Meetings
- Analytical techniques

[Planning]

9 QUESTION

How is the issue log used
in control stakeholder
engagement?

10 QUESTION

What is the key benefit
of manage stakeholder
engagement?

ANSWER

As an input because it is updated as new issues are identified, and current issues are completed.

[Monitoring and Controlling]

ANSWER

It allows the project manager to increase support and minimize resistance from stakeholders, which significantly increases the chances to achieve project success.

[Executing]

11

List three enterprise environmental factors that may influence identify stakeholders.

12

List five items from the project plan useful in plan stakeholder management.

ANSWER

- Organizational culture and structure
- Government or industry standards such as regulations or product standards
- Global, regional, or local trends; practices; and habits

[Initiating]

ANSWER

- Life cycle for the project and the processes in each phase
- Description of the work to be done to accomplish the project objectives
- Description of how human resource requirements will be met and how roles and responsibilities, reporting requirements, and staffing management will be addressed and structured
- Change management plan
- Needs and techniques for stakeholder communications

[Planning]

13

QUESTION

Why is project governance
critical to manage
stakeholder expectations?
List two examples.

14

QUESTION

Why is stakeholder
management more than
improving communications
and requires more than
managing a team?

ANSWER

It serves to align the project with stakeholder needs and objectives.

It provides a framework for the project manager and sponsors to make decisions to satisfy stakeholder needs and expectations and organizational strategic objectives or address situations where there is not alignment.

[Executing]

ANSWER

It is about creating and maintaining relationships between the project team and its stakeholders in order to satisfy their respective needs and requirements within project boundaries.

[Planning]

15

Why is identify stakeholders continuous throughout the project life cycle?

16

Provide five examples of project documents that may be useful for control stakeholder engagement.

ANSWER

Stakeholders have varying levels of responsibility and authority on projects, which may change over the life cycle, and different stakeholders will be identified as the project continues.

[Initiating and Executing]

ANSWER

- Project schedule
- Stakeholder register
- Issue log
- Change log
- Project communications

[Monitoring and Controlling]

17

When is the ability of a stakeholder to influence the project the highest?

18

Define Plan Stakeholder Management.

ANSWER

During the initial stages of the project, and it becomes progressively lower as the project progresses.

[Initiating, Planning, Executing, and Monitoring and Controlling]

ANSWER

Developing appropriate management strategies to effectively engage stakeholders in the project life cycle based on analyzing their needs, interests, and potential impact on project success.

[Planning]

19

When should the project
manager use expert
judgment in plan stakeholder
management?

20

List five items from the
communications management
plan that can provide
guidance on manage
stakeholder expectations.

ANSWER

Based on the project objectives to decide upon the level of engagement required at each stage of the project from each stage and to help create the stakeholder management plan.

[Planning]

ANSWER

- Stakeholder communications requirements

- Information to be communicated including language, format, content, and level of detail

- Reason to distribute information

- Person or group to receive the information

- Escalation process

[Executing]

21

Provide four examples as to why identify stakeholders is critical.

PROJECT **Stakeholder** MANAGEMENT

22

List six main categories of stakeholders.

PROJECT **Stakeholder** MANAGEMENT

ANSWER

Failure to do so can lead to:

- Delays

- Cost increases

- Unexpected issues

- Other negative consequences including project cancellation

[Initiating, Planning, Executing, and Monitoring and Controlling]

ANSWER

- Sponsors

- Customers or end users

- Sellers

- Business partners

- Organizational groups

- Functional managers

[Initiating]

23

Provide four examples of other stakeholders that may have a financial interest in the project, contribute inputs to it, or have an interest in the project's outcome.

PROJECT **Stakeholder** MANAGEMENT

24

What are four organizational process assets considered in manage stakeholder engagement?

PROJECT **Stakeholder** MANAGEMENT

ANSWER

- Financial institutions
- Government regulatory agencies
- Subject matter experts
- Consultants

[Initiating]

ANSWER

- Organizational communication requirements
- Issue management procedures
- Change control procedures
- Historical information about previous projects

[Executing]

25

What are six organizational process assets that may require updates as a result of control stakeholder engagement?

PROJECT **Stakeholder** MANAGEMENT

26

Why is it important to manage stakeholder expectations through negotiation and communications?

PROJECT **Stakeholder** MANAGEMENT

- Stakeholder notifications
- Project reports
- Project presentations
- Project records
- Feedback from stakeholders
- Lessons learned documentation

[Monitoring and Controlling]

PROJECT **Stakeholder** MANAGEMENT

To ensure project goals are achieved.

[Executing]

PROJECT **Stakeholder** MANAGEMENT

27

QUESTION

Define manage stakeholder
engagement.

28

QUESTION

List five ways to classify
the level of engagement of
stakeholders.

ANSWER

The process of communicating and working with stakeholders to meet their needs/expectations, address issues if they occur, and foster stakeholder engagement in project activities during the life cycle as appropriate.

[Executing]

ANSWER

- Unaware
- Resistant
- Neutral
- Supportive
- Leading

[Planning]

29

How can manage stakeholder expectations increase the probability of project success?

30

List three organizational process assets that can influence the identify stakeholders process.

ANSWER

By ensuring stakeholders clearly understand the project's goals, objectives, benefits, and risks so they are active project supporters and can help guide activities and decisions.

[Executing]

ANSWER

- Stakeholder register templates

- Lessons learned from previous projects or phases

- Stakeholder registers from previous projects

[Initiating]

31

List four ways the stakeholder management plan may be used in manage stakeholder engagement.

32

List two project documents that may require updates because of the control stakeholder engagement process.

ANSWER

- To provide guidance on how to best involve the stakeholders in the project

- To provide methods and technologies for stakeholder communications

- To determine the level of interaction among project stakeholders

- To help define a strategy to identify and manage stakeholders throughout the project life cycle

[Executing]

ANSWER

Stakeholder register

Issue log

PROJECT **Stakeholder** MANAGEMENT

PROJECT **Stakeholder** MANAGEMENT

33

What is Project Stakeholder
Management?

PROJECT Stakeholder MANAGEMENT

34

Define control stakeholder
engagement.

PROJECT Stakeholder MANAGEMENT

ANSWER

The processes required to identify the people, groups, or organizations that could be impacted by the project, to analyze stakeholder expectations and their impact on the project, and to develop appropriate strategies to effectively engage them in project decisions and outcomes.

[Initiating, Planning, Executing, and Monitoring and Controlling]

ANSWER

The process of monitoring overall project stakeholder relationships and adjusting plans and strategies as needed to engage stakeholders.

[Monitoring and Controlling]

35

List eight items to include in the stakeholder management plan.

36

List five seven examples of work performance data used in control stakeholder engagement.

PROJECT **Stakeholder** MANAGEMENT

PROJECT **Stakeholder** MANAGEMENT

ANSWER

- Desired and current engagement levels of key stakeholders
- Scope and impact of change to stakeholders
- Identified interrelationships and potential overlap between stakeholders
- Stakeholder communication requirements for the current project phase
- Information to be distributed to stakeholders
- Reason to distribute the information and expected impact on stakeholder engagement
- Time frame and frequency to distribute the required information to stakeholders
- Method for updating and refining the plan during the project

[Planning]

ANSWER

- Reported percentage of work completed
- Technical performance measures
- Start and finish dates of schedule activities
- Number of change requests
- Number of defects
- Actual costs
- Actual durations

[Monitoring and Controlling]

PROJECT **Stakeholder** MANAGEMENT

PROJECT **Stakeholder** MANAGEMENT

37 QUESTION

What is true about every
project?

38 QUESTION

Define stakeholder analysis.
Where is it used?

ANSWER

It will have stakeholders who are impacted by or can impact the project in a positive or negative way.

[Initiating, Planning, Executing, and Monitoring and Controlling]

PROJECT **Stakeholder** MANAGEMENT

ANSWER

A technique to systematically gather and analyze quantitative and qualitative information to determine whose interests should be taken into account during the project.

It is a tool and technique in identify stakeholders.

[Initiating]

PROJECT **Stakeholder** MANAGEMENT

39

Eleven subsidiary plans to the project management plan may need to be updated as a result of the control stakeholder engagement process. List them.

40

What is the key benefit of the identify stakeholder process?

ANSWER

- Change management plan
- Communications management plan
- Cost management plan
- Human resource management plan
- Procurement management plan
- Quality management plan
- Requirements management plan
- Risk management plan
- Scope management plan
- Schedule management plan
- Stakeholder management plan

[Monitoring and Controlling]

PROJECT **Stakeholder** MANAGEMENT

ANSWER

It allows the project manager to identify the appropriate focus for each stakeholder or group of stakeholders.

[Initiating]

PROJECT **Stakeholder** MANAGEMENT

41

What is meant by a leading stakeholder?

42

List the four inputs to identify stakeholders.

PROJECT **Stakeholder** MANAGEMENT

PROJECT **Stakeholder** MANAGEMENT

ANSWER

A stakeholder that is aware of the project and its potential impacts and is actively engaged in ensuring the project is a success

[Planning]

ANSWER

- Project charter
- Procurement documents
- Enterprise environmental factors
- Organizational process assets

[Initiating]

43

QUESTION

List the three steps that are followed during a stakeholder analysis.

44

QUESTION

What is meant by a supportive stakeholder?

ANSWER

- Identify all potential project stakeholders and relevant information about each one

- Analyze the potential impact or support each stakeholder could generate and classify them

- Assess how key stakeholders are likely to respond or react in various situations

[Initiating]

ANSWER

A stakeholder that is aware of the project and its potential impacts and is supportive of change

[Planning]

45

QUESTION

How can the project manager best focus on the relationships necessary to ensure project success?

46

QUESTION

What are two examples of change requests from control stakeholder engagement?

ANSWER

By classifying stakeholders according to their interest, influence, and involvement in the project recognizing the influence of a stakeholder may not occur or become evident until later stages in the project or phase.

[Initiating]

ANSWER

- Recommended corrective actions to bring expected future performance of the project in line with the project management plan

- Recommended preventive actions to reduce the possibility of incurring future negative project performance

[Monitoring and Controlling]

47

What are the four classification models used in stakeholder analysis?

48

What is meant by a resistant stakeholder?

ANSWER

- Power/interest grid
- Power/influence grid
- Influence/interest grid
- Salience model

[Initiating]

ANSWER

A stakeholder that is aware of the project and its potential impacts but is resistant to change.

[Planning]

49

List seven categories of people or groups who could provide expert judgment in identify stakeholders.

50

How are profile analysis meetings used?

ANSWER

- Senior managers
- Other units in the organization
- Identified key stakeholders
- Project managers who manage projects in the same area
- Subject matter experts in the business or project area
- Industry groups or consultants
- Professional and technical associations, regulatory bodies, and non-governmental organizations

[Initiating]

ANSWER

As a tool and technique in identify stakeholders to develop an understanding of major project stakeholders and to exchange information about roles, interests, knowledge, and the stakeholder's overall position about the project.

[Initiating]

51

QUESTION

What is the main output of
identify stakeholders? What
are its three main categories?

52

QUESTION

Provide an example of
the sensitive nature of the
stakeholder management
plan.

ANSWER

- The stakeholder register

Categories:

- Identification information
- Assessment information
- Stakeholder classification

[Initiating]

ANSWER

Information on stakeholders who may be resistant to the project will be in the plan and can be potentially damaging if they receive it. The project manager must consider carefully the recipients of this plan.

[Planning]

53

QUESTION

What should be reviewed
each time the stakeholder
management plan is updated?

54

QUESTION

What are the four key activities
in manage stakeholder
expectations?

ANSWER

The validity of underlying assumptions to ensure continued accuracy and relevancy.

[Planning]

ANSWER

- Engaging stakeholders at appropriate project stages

- Managing stakeholder expectations through negotiation and communication

- Addressing potential concerns that have not become issues and anticipating future problems that may be raised to stakeholders

- Clarifying and resolving identified issues

[Executing]

55

QUESTION

List six organizational process assets to update as an output of manage stakeholder engagement.

56

QUESTION

What is the key benefit of the manage stakeholder engagement process?

ANSWER

- Stakeholder notifications
- Project reports
- Project presentations
- Project records
- Feedback from stakeholders
- Lessons learned documentation

[Executing]

ANSWER

It will maintain or increase the effectiveness and efficiency of stakeholder engagement activities as the project evolves and its environment changes.

[Monitoring and Controlling]

PROJECT **Stakeholder** MANAGEMENT

PROJECT **Stakeholder** MANAGEMENT

57

What are four key interpersonal skills the project manager can use to manage stakeholder's' expectations?

58

Name the two project documents to update as an output of plan stakeholder management.

ANSWER

- Building trust
- Revolving conflict
- Active listening
- Overcoming resistance to change

[Executing]

ANSWER

- Project schedule
- Stakeholder register

[Planning]

59

Name four general
management skills the
project manager can use to
coordinate and harmonize the
group toward accomplishing
project objectives.

PROJECT **Stakeholder** MANAGEMENT

60

What are three examples of
performance information
as an output from control
stakeholder expectations?

PROJECT **Stakeholder** MANAGEMENT

ANSWER

- Facilitate consensus toward project objectives

- Influence people to support the project

- Negotiate agreements to satisfy the project needs

- Modify organizational behavior to accept the project outcomes

[Executing]

ANSWER

- Status of deliverables

- Status of change requests

- Forecasted estimates to complete

[Monitoring and Controlling]

Printed and bound by CPI Group (UK) Ltd, Croydon, CR0 4YY

25/10/2024

0177935354-0001